Heat, Greed and Human Need

For Anna

And for Amy, George, Oliver, Ted, Esther and Will

Heat, Greed and Human Need

Climate Change, Capitalism and Sustainable Wellbeing

Ian Gough

Visiting Professor, Centre for the Analysis of Social Exclusion, London School of Economics, UK

 Edward Elgar
PUBLISHING

Cheltenham, UK • Northampton, MA, USA

Published by
Edward Elgar Publishing Limited
The Lypiatts
15 Lansdown Road
Cheltenham
Glos GL50 2JA
UK

Edward Elgar Publishing, Inc.
William Pratt House
9 Dewey Court
Northampton
Massachusetts 01060
USA

A catalogue record for this book
is available from the British Library

Library of Congress Control Number: 2017941901

This book is available electronically in the **Elgar**online
Social and Political Science subject collection
DOI 10.4337/9781785365119

ISBN 978 1 78536 510 2 (cased)
ISBN 978 1 78536 512 6 (paperback)
ISBN 978 1 78536 511 9 (eBook)

Typeset by Servis Filmsetting Ltd, Stockport, Cheshire
Printed and bound by CPI Group (UK) Ltd, Croydon, CR0 4YY

Contents

Figures

Tables

Acknowledgements

Thanks are due first of all to Graham Room at the University of Bath, with whom I discussed the relationship of climate change and social policy a decade ago and who encouraged me to edit a debate on the topic that appeared in the *Journal of European Social Policy* that he was co-editing. John Hills and David Held supported my visiting professorship at the LSE, and John and colleagues in the Centre for the Analysis of Social Exclusion (CASE) provided a welcoming base when I moved to London in 2009. At the same time the ESRC awarded me a two-year Small Grant to research 'Climate change and social policy: Rethinking the political economy of the welfare state' (ES/H00520X/1). Subsequently, Simon Dietz and the Grantham Research Institute on Climate Change and the Environment at the LSE have funded generous research assistance. This afforded me the excellent research and bibliographic skills of in turn Sam Marden, Cindy Smith, Alex Stark, Erin Nash and Geraldine Satre Buisson. James Angel at the New Economics Foundation and Mario Battaglini at CASE have also provided timely research help. Thanks too to Alex Pettifer at Edward Elgar for his enthusiastic support for this book.

For reading and commenting on the entire first draft I'm deeply grateful and indebted to John Barry, Kate Pickett, Graham Room and Robin Stott. Other colleagues and friends who've advised on chapters or substantial chunks of the manuscript include Tania Burchardt, Simon Dietz, Fergus Green, Geoff Hodgson, Giorgos Kallis, Tim Kasser, Max Koch, Stephan Leibfried, Jane Millar, John O'Neill, Narasimha Rao, Julia Steinberger and Marko Ulvila. I'm grateful to them all. Since they all disagree with each other in many ways it is perhaps unnecessary to state that I alone am responsible for the final product.

This book is based on a dozen or more articles and chapters written over the past decade, and these too have been guided and assisted by other colleagues and friends – as well as a collection of anonymous referees. They include Alex Bowen, Sarah Cook, Michael Dover, Len Doyal, Robert Falkner, Kevin Farnsworth, Des Gasper, Howard Glennerster, Monica Guillen-Royo, John Hills, Michael Jacobs, Alexandra Kaasch, James Meadowcroft, Paul Ormerod, Guy Standing, Paul Stubbs, Peter

Taylor-Gooby, Göran Therborn, Karen Turner and Polly Vizard. Peter Taylor-Gooby also contributed to the title.

Google and the Mac Finder have progressively replaced my waning memory. Without them this book would have taken much longer to write.

Finally, this book would not exist at all were it not for Anna Coote. She it was who convinced me that climate change must be taken seriously. Having achieved that, she has been with me from the very start of the project to this end. Her work on the Sustainable Development Commission and subsequently at the New Economics Foundation has been an inspiration. She has supplied ideas and arguments, maintained my optimism, kept me focused on longer-term radical futures, yet insisted on linking ideas with policy and implementation, and cajoled me to write in reasonably accessible English. This book would not exist without her love and collaboration, and I dedicate it to her.

Introduction

THE RATIONALE FOR THIS BOOK

Climate change – the 'heat' of the title – is the most encompassing and threatening of all the planetary limits that characterise this new age of the Anthropocene. It is subject to a wide range of research and analysis. My hope is that this book adds five things to the current literature.

First, I advance a new concept with which to grasp the social impact of climate change – that of human need. Climate change threatens human wellbeing across the globe today and into the future. To address this requires a measure that is constant across both space and time. I argue that the only candidate is basic human needs and the extent to which they are satisfied. Consumer preferences will not do, nor will 'happiness': since both are affected by present circumstances and institutions, they lead us into a circular process which can neither escape the driving forces of the past nor encompass the future. A theory of human need is developed that provides a universalist framework and at the same time enables what I call 'need satisfiers' to be identified in a myriad of specific circumstances, across different contexts and cultures. This provides the normative and ethical underpinning for evaluating the social dimension of climate change.

Second, I try to overcome the too frequent gulf between idealist visions of a different world and hard-headed appreciation of the current global system. So this book provides an economic, social and political analysis of the drivers of climate change. At the core lies capitalism – the 'greed' of the title – and the relentless processes of accumulation, growth and inequality that it reproduces. Alongside this is the global system of nation states. Together they mobilise the interests and institutions of the modern world and shape our dominant ideas. Any attempt to halt and reverse global warming will have to begin within this encasement. The book argues for a clear-headed appreciation of current political economy.

Third, my approach is multi-disciplinary. Scholarship on climate change has been dominated by natural scientists and economists, studying the environment and 'the economy' and their interactions. The social dimension is much less developed. It encompasses vital issues such as equity, justice, inequality, poverty and empowerment but often with little coherence. It is essential to bring in other social sciences to give credence to the truth that we live social lives within structures of power, both overt and hidden. Multi-disciplinary research provides an essential antidote to the continuing prevalence of neoclassical economics and the assumptions of neoliberalism. But it inevitably spreads the net wide rather than deep: my hope is that the synergies derived from a broad reach will more than compensate for the absence of disciplinary focus.

Fourth, this book stems from a life-long interest in social policy – the mobilisation of collective action and state power to improve human wellbeing. Yet, with a few exceptions, the study of social policy has (blindly or wilfully) ignored the environment and the planetary limits within which the pursuit of human needs and wellbeing must necessarily take place. The second part of this book attempts to close that gap in some detail. It analyses the 'welfare states' of the developed world: how far they are dependent on the carbon economy and how they can be reformed to pursue simultaneously both carbon mitigation and human welfare. This leads into analyses of policy-making under different scenarios of production, consumption and growth and proposes certain 'eco-social' policies that could combine sustainable livelihoods with human wellbeing.

Finally, the book concludes that the strategy of 'green growth', which underlies the Paris 2015 agreement and will dominate in the coming years, will only work if it is treated as a stepping stone to a political economy based on needs, sufficiency and redistribution, not on continuing economic growth. Green growth alone will not be enough, yet 'degrowth' seems impossibly daunting in political terms. Transitional strategies are needed to move from the former to the latter. The book develops an intermediate stage where consumption in rich countries is 'recomposed' away from high-carbon luxuries to low-carbon necessities. This three-stage process – from green growth, through recomposed consumption to degrowth – seems to me the only way to progress from the hard-headed 'greed' and technological might of contemporary capitalism to an ethical, just and sustainable future.

THE CONCEPTUAL FRAMEWORK

To use an old distinction, this book engages – perhaps recklessly – with both normative and positive issues. *Normative* statements make claims about how things should be and how to value them; they entail a standard, rule or principle used to judge or direct human conduct. *Positive* statements purport to describe and/or interpret events in the 'real world'.

These are simplified definitions,[1] but they help map out the distinctive approaches of need theory and political economy in this book.

The Normative Framework: Human Need

The terms *welfare* and *wellbeing*[2] raise a host of questions. The dominant conception today is that of welfare economics, which is primarily concerned with utility or the satisfaction of consumer preferences (now closely pursued by a school pressing the claims of 'happiness'). This rests on two normative foundations: that individuals are the best judges of their own preferences or wants, and that what is consumed should be determined by the private consumption preferences of individuals. It precludes questioning the nature and content of consumer preferences, except within narrow limits. However, it has been subject to numerous challenges, on the grounds of subjectivity, epistemic irrationality, endogenous and adaptive preferences, the limitlessness of wants, the absence of moral evaluation, and the non-specificity of future preferences (Gough 2015a).

An alternative is required, and Chapter 2 summarises *A Theory of Human Need* (Doyal and Gough 1991) and related frameworks. The essential premise is that all individuals, everywhere in the world, at all times present and future, have certain basic needs. These must be met in order for people to avoid harm, to participate in society and to reflect critically upon the conditions in which they find themselves. Only if we understand needs in this way – in universal terms, applied across time and place – can we plan for and measure progress towards our social and environmental goals, both globally and into the future. Having identified this core objective, I go on to address the immense cultural variety in ways of meeting needs and set out a methodology for identifying need satisfiers in particular social settings.

According to this and allied theories, human needs are objective, plural, non-substitutable and satiable. This means that needs are also *cross-generational* – a crucial point since global warming will progressively impose dilemmas of intergenerational equity. We can assert with much confidence that the basic needs of future generations of humans will be the same as those of present generations. Moreover, human needs, unlike preferences, have a sound ethical grounding: they come with claims of justice and equity in tow. Universal needs imply ethical obligations on

---

[1] For example Putnam has written on 'entangled concepts' that are both evaluative and descriptive (Putnam 2002).
[2] Throughout this book these terms are used interchangeably.

the part of individuals as well as claims of justice – universal rights and obligations – on social institutions. An important corollary is that meeting human needs should be given priority over meeting wants if the two conflict or if resources are scarce. Human needs, present and future, trump present (and future) consumer preferences.

The Positive Framework: Political Economy

A major part of the book addresses 'positive' issues, describing and explaining features of what we can loosely call 'climate capitalism'. It utilises political economy but also draws on ecology and social theory to develop what might be called an *eco-social political economy* approach (ESPE). This aims to bring into a more embracing framework the following elements: the economy (profits and the drive to accumulate capital), ecology (ecosystems and organisms, including humans), the social domain (paid and unpaid labour, human and social resources and relationships, and inequality) and the political (states, governance and power). The intention is to provide a realist framework for analysing the contemporary drivers and consequences of climate change. I discuss in turn: the relationship between environment, society and economy; capitalism as a system; the ecological and social domains; and the neoliberal era.

Environment–society–economy

The domain of the *economy*, which is central to both the environmental and the social domains, refers to the monetised commodity economy made near-universal in the latest phase of globalisation. It is one of the three interrelated domains of sustainable development, along with society and environment. Chapter 1 presents Kate Raworth's 'doughnut' or 'lifebelt', which understands that the biophysical environment frames societies and economies, and that the social dimension establishes the values and components of need satisfaction or sustainable wellbeing. It implies that the economy is a means to these ends. 'The economy is, in the first instance, a subsystem of human society . . . which is itself, in the second instance, a subsystem of the totality of life on Earth (the biosphere). And no subsystem can expand beyond the capacity of the total system of which it is a part' (Porritt 2006: 46). This idea of nested domains can be pictured as a series of concentric circles, with the biosphere encompassing society, which in turn encompasses the economy.

However, the global capitalist economy is a mighty system with an inbuilt dynamic propelled by actors with immense power, so much so that it is already overwhelming planetary boundaries. In practice, far from being the junior domain, it dominates. To visualise the interrelations

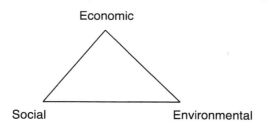

Figure I.1   The sustainable development triangle

between the three domains a Venn diagram or triangle can also be useful (Figure I.1).

The three domains or points of the triangle are traditionally associated with clusters of different values, goals and policy domains:

Economy	*Society*	*Environment*
Efficiency	Human needs	Ecology
Productivity	Equity/equality	Biophysical system
Competitiveness	Wellbeing	Planetary limits
Economic growth	Social justice	Ecosystem services
	Social inclusion	Intergenerational justice
Economic policy	*Social policy*	*Environmental policy*

The downside of this visual metaphor is that the economic, environmental and social clusters look both separate and equal. They are neither, as this book makes clear.

The ESPE approach is multi-disciplinary, drawing selectively on analytical perspectives from social sciences – notably environmental economics, ecological economics, Marxism, sociology, political science, historical institutionalism, and social policy. I make no claims for ESPE as a theory, merely as a useful framework to keep in mind the twin encasements of the 'economy' – the social and the biophysical. Yet to speak of, and write for, the 'social sciences' today is a daunting task. Academic disciplines rule the roost, and research that crosses their boundaries, though constantly advocated, is a struggle to realise in practice. In particular there is a gulf of incomprehension and misunderstanding between economics and other social science disciplines.

The dominant paradigm framing analysis of the economy over the past four decades has been neoclassical economics. I would claim that, as a positive analytical approach, this suffers from at least three blind spots. First, it equates (uninsurable) uncertainty with (insurable) risk, despite the looming dangers of climate tipping-points and catastrophic outcomes, and thereby many forms of cost–benefit analysis of the future impacts of climate change mitigation. Second, it assumes that technical progress is 'exogenous' – determined outside the economic system model – which impairs the ability to analyse the big, non-marginal changes required by rapid economic transformation and decarbonisation and renders all current economic models unviable according to some economists. Third, standard economics ignores or underestimates the role of historical and institutional barriers in implementing effective climate policies (Dietz 2011; Hodgson 2013; Scrieciu et al. 2013; Grubb et al. 2014; Farmer et al. 2015).

A survey in 2011 concluded that neoclassical economics advocated a 'modest' optimal rate of emissions reduction, well below that recommended by climate scientists (Dietz 2011). Economic rationales for deeper emission cuts can be, and have been, justified by using a lower discount rate to estimate future climate costs (Stern 2007), or by recognising the existence of low-probability, high-impact scenarios (Weitzman 2009). It is probably true to say that neoclassical climate economists are shifting towards a more precautionary approach, but they still lag well behind the accelerating warnings of climate scientists. A variety of economic concepts have value and are used elsewhere, for example in Chapter 6.

Political economy provides a more encompassing and fruitful framework. It embraces two core assumptions (Caporaso and Levine 1992; Gamble 1995). The first is that political and economic processes, though analytically distinct under capitalism, are interlinked and should be studied as a complex and interrelated whole. The second is that the economy, the sphere of 'material provisioning', has a special weight in explaining and properly understanding the polity and politics.[3] Governments are not perceived as neutral umpires correcting malfunctions in the market economy, but as central institutions both reflecting and shaping the distribution of power and resources in society. The political economy approach can be found in both Marxist and non-Marxist guises, the latter for example in Lindblom (1977), Strange (1988), Dahl (1998) and Hacker and Pierson (2002).

Capitalism as a system

Political economy is willing to refer to *capitalism* and the capitalist system, a now global system driving the relationship between economy, climate change and human needs. Though the exact definition of capitalism is

---

[3]　There are two major schools using the label 'political economy'. Public choice theory, grounded in the assumptions and methods of neoclassical economics, is not considered here.

disputed, there is agreement on some key features. The first is the production of commodities for profit. As first noted by Karl Marx, capitalism subordinates the production of 'use-values' to meet human wants and needs[4] to a drive to accumulate 'exchange value' by producing commodities for sale:

> The simple circulation of commodities – selling in order to buy – is a means of carrying out a purpose unconnected with circulation, namely the appropriation of use-values, the satisfaction of wants. The circulation of money as capital is, on the contrary, an end in itself, for the expansion of value takes place only within this constantly renewed movement. The circulation of capital has therefore no limits. Thus the conscious representative of this movement becomes a capitalist. His person, or rather his pocket, is the point from which the money starts and to which it returns . . . The restless never-ending process of profit-making alone is what he aims at. (Marx 1926: 169, 171)

This applies whether this thing called capital is invested in energy, industry, agriculture, services, distribution, finance or the immaterial digital world. This is the central driving force of the capitalist world economy.

Another key feature of capitalism is the private ownership of means of production or capital (or at the least a substantial privately owned sector). Private firms therefore control investment in means of production and play a major role in determining economic paths of development. A further feature is the existence of a class of people who sell their labour-power for wages; they are without property of their own, or at least without sufficient property to support themselves and their families over their lifetime. Both these features emphasise the role of property rights, or their lack, and legal institutionalists would argue that a system of law recognising legal rights over many kinds of asset is another fundamental component of capitalism (Hodgson 2016).

Capitalism is a system that has evolved in historical time, from mercantile beginnings in north-western Europe in the sixteenth century, through industrial capitalism in Britain in the eighteenth century and it spread to Europe, North America and other settler countries in the nineteenth century. In the twentieth century new socio-technical forms of 'Fordism' – integrated production and mass consumption – emerged initially in the US and spread across the advanced capitalist world. Since the 1970s this has morphed into the global and financialised system that characterises the new millennium. As Marx predicted, the system has spread both intensively, commodifying many of our activities and relationships, and extensively, 'filling up' the entire world. The pursuit of profit harnesses,

---

[4] The distinction between the two can be ignored at this stage of the argument.

shapes and drives technological progress. The result is the accelerating transformations of our economies and lives – and of the globe. The goal of endless economic growth is a necessary corollary of capitalism.

Of epochal importance has been the coevolution of capitalism around fossil hydrocarbons. Since the late eighteenth century fossil fuels have provided the indispensable energy basis for this process of accumulation. With the burning of coal, energy was freed from wind, water, biofuels and animal power, and millions of years of 'stored sunlight' energy suddenly became available. The later exploitation of oil provided even more con-centrated sources of energy. The fundamental driver of global warming has been a combination of fossil-based industrialisation and global capitalism – carboniferous capitalism (Newell and Paterson 2010).

The ecological and social domains
Societies have always evolved mechanisms to protect and reproduce labour and nature. Activities to care for and socialise children, build and maintain communities and create shared meanings have always existed outside the market economy and mainly been undertaken by women, as feminist analysts stress (Elson 1988; Mellor 1997; Barry 2012; Fraser 2014). Similarly, forms of collective action have evolved throughout history to counteract the over-exploitation of environmental 'common pool resources' in local contexts, such as overfishing or overuse of water supplies, as Elinor Ostrom (1990) has documented. These take numerous forms, such as communal tenure or rights to environmental services, with voluntary and communal policing and enforcement.

The reproduction of human capacities and environmental services in local contexts has for long depended on pre-capitalist, uncommodified and collective arrangements. But the supply of natural and human resources cannot be guaranteed from within the capitalist dynamic. Both sets of resources require protection delivered by institutions with distinctive values and driving forces: what Nancy Fraser (2014) calls 'the hidden abodes of capitalism'. If the dominant value of the capitalist system is 'legitimate greed', the values of the 'hidden abodes' are social mores such as solidarity, mutual support, and communal and collective responsibility, which over time have mutated into ideas and movements for social citizenship and eco-logical sustainability (cf. Streeck 2014). Both nature and human needs are complex, messy and multi-dimensional, yet a capitalist economy is ultim-ately dependent on the reproduction of both (Koch 2012).

Capitalism as a mode of production driven by accumulation is always encased within particular social formations, comprising social and polit-ical institutions. Karl Polanyi's concept of the 'double movement' in *The Great Transformation* provides a useful historical perspective. As the market economy in nineteenth-century Britain became self-regulating and 'disembedded' from social institutions and social patterns of behaviour, a contradiction emerged in the social domain. Aside from slavery, labour itself cannot be commodified: it is a 'fictitious commodity' because it is not produced for sale and 'cannot be detached from the rest of life' (Polanyi 1944). Its de facto commodification resulted in profound insecurity and threats to welfare, which led to counter-mobilisations by workers, communities and social reformers and the political rise of the 'social question'. New forms of regulation and protective institutions, including social policies, emerged – piecemeal and in a wide variety of forms – to cope with the unplanned, harmful and system-threatening effects of the commodification of labour. After the Second World War this societal reaction was institutionalised in the form of 'welfare states'.

The Polanyian perspective can also be applied to understand state interventions in the environmental domain and in climate policy. Nature is – like labour – a 'fictitious commodity'. Early industrialisation set in train the inexorable rise in CO_2 emissions and accelerated the commodification of natural resources and the despoliation of the environment. This in turn stimulated early counter-movements in the nineteenth century such as municipal regulations over water supply. This was renewed in the late twentieth century when the threats of unregulated industrial and consumption growth motivated environmental movements and green politics. Again this gradually pressured state interventions and environmental governance, including eventually early measures to restrain greenhouse gas (GHG) emissions. Some political scientists have interpreted the emergence of 'eco-states' as a parallel to the earlier emergence of welfare states (Barry and Eckersley 2005; Meadowcroft 2005; Gough et al. 2008a; Duit 2014; Duit et al. 2016: 3; cf. Mol 2016).

But the compatibility of capitalism with ecological constraints is even more problematic. *Ecological* economics recognises the ontological and normative priority of a sustainable biosphere but goes on to analyse the numerous interactions that arise. Key themes include scale, coevolution, uncertainty and complexity (Özkaynak et al. 2012). Ecosystems can be disturbed when the *scale* of the human economy grows abnormally in relation to its environment. A central question is that of limits – brought to international attention by the publication of *Limits to Growth* (Meadows et al. 1972) – and the ability of technology to circumvent such limits (Costanza 1991). Herman Daly (1996) argues that there are fundamental limits to economic growth stemming from 'finitude, entropy and ecological interdependence' and that technological progress or 'ingenuity' cannot endlessly shift outwards the boundaries of the possible: humanity has already moved from an 'empty' to a 'full' world (Daly 2007). Or, in Boulding's (1966) telling language, we are moving from a 'cowboy economy' lived on limitless plains to a 'spaceman economy' lived within a capsule.[5] These issues are revisited in Chapter 8.

From this perspective, capitalism depends on both the ecological sphere and the social sphere: without collective regulation of both, its own survival is threatened. Other disciplines and fields of study can augment our understanding of these mechanisms. Modern-day *institutionalism* is a prominent paradigm in fields of study including economics, political science, international relations and social policy. Institutions can be most broadly defined as durable systems of established and prevalent social rules that structure social interactions. Repeated over time such behaviour acquires normative weight (Hodgson 2006). Systems of government are of course central institutions, both reflecting and shaping the distribution of power and resources within societies. For example, Chapter 5 builds on welfare state scholarship, comprising both historical and comparative cross-national research with a rich variety of methodologies (Castles et al. 2010). As a way of summarising shifts in state policies and their drivers I utilise an earlier framework (Gough 2008, 2016). This posits five drivers of policy development: the 'five Is' of industrialisation (and other structural trends), interests, institutions, ideas/ideologies, and international influences.

Finally, *co-evolutionary* perspectives recognise the conjoint and complex development of natural and social systems, as in new research in the sociology of science, technology and markets and the political science of socio-ecological or green transitions (IPCC 2014c).

The neoliberal era

Post-war capitalism was often characterised as Keynesian welfare state capitalism. A new social settlement or class compromise re-established private ownership of property and markets but contained within a public commitment to full employment, state regulation of key industries and a comprehensive welfare state. During the 1970s this mode of regulation began to be replaced by a very different form frequently designated as *neoliberal capitalism* or *neoliberalism*. This term must be used with care, since it is associated with a bewildering range of meanings (Venugopal

---

[5]   On the other hand, the potential for solar- and gravity-powered energy to replace carbon fuels (though with less efficiency) is evident and is almost limitless. This does not question the second law of thermodynamics, that the universe is constantly becoming more entropic or disordered, its energy and matter less available for use. It recognises that the Earth is on the contrary an open system constantly powered by a flow of energy from the sun. This enables the flow of entropy within the biosphere to be reversed and with better husbandry can reduce entropy due to human activity (Jacobs 1991).

2015). I take 'actually existing neoliberalism' to be a distinct phase of capitalism since around 1980. It embraces a dominant set of ideas and of practices. Its defining ideas included a belief in the superiority of markets and a denigration of much government and collective action. Its defining characteristics include a new international division of labour, the global spread of production networks, trade and financial flows, the dominance of finance, rising profit shares and widening inequalities within countries (Glyn 2006; Newell and Paterson 2010; Duménil and Lévy 2011; Koch 2012; Stiglitz 2013).

In the new era the relative power of business corporations and the financial sector has grown, especially relative to trade unions and labour interests, but also vis-à-vis nation states. This stems not only from their lobbying power but also from their structural power, the ability to influence policy without having to apply direct pressure on governments through agents. This stems from several factors, but two have become more important in recent decades: the ability to shift investment and economic activity between jurisdictions and the structural position of finance capital in ensuring national economic survival. The end result is a closer symbiosis or even 'capture' of governments by big business and finance (Gough 2000, ch. 4; Hacker and Pierson 2002; Woll 2014). It is a historical coincidence (and, I would say, a tragedy) that the widespread recognition of climate change as a global threat has arisen in the neoliberal era – what Naomi Klein (2015) calls the great misfortune of Bad Timing.

To conclude, *eco-social political economy* is used here as shorthand for an approach that draws on the contributions that social science disciplines can make to comprehending the impacts of global warming on human action, social systems and human wellbeing – and vice versa. It recognises that proposals and movements to change direction are situated within historical time and in a context of social and cultural practices, which calls for cross-disciplinary study. The eco-social approach enables us to scrutinise the interrelationships and three-way conflicts between the domains of the biosphere (notably global warming), society and the economy. This is the terrain on which scientific knowledge and public concern over climate change have taken root and grown. Eco-social political economy provides the main analytical framework for this book.

PLAN OF THE BOOK

The book is divided into two parts: the first covering conceptual and global issues, the second the affluent world. My original idea was to study climate change, inequality and social policy within the rich world of the

North, and that is the aim of Part II. But climate change is *the* global threat posing existential dangers while at the same time posing wicked dilemmas in coordinating global action to constrain it. These issues are of epochal significance in their own right, which is justification enough for Part I. They also frame the responsibilities and obligations of the North towards the planet and the peoples of East and South. The basic assumption here is that speedy 'contraction and convergence' must take place. Thus there is a double obligation on the rich world to decarbonise rapidly its production and consumption practices and to fund generously mitigation and adaptation programmes in the global South. This is the fundamental assumption framing the discussion of economic and social issues and policies in Part II of the book.

Chapter 1 summarises our best knowledge about the predicted future of global warming and its potentially catastrophic implications for human habitats and human wellbeing. The policy options are summarised, divided between programmes to mitigate climate change and to adapt to it. But climate policy alone could be unjust and inequitable. The goal must be to respect biophysical boundaries while at the same time pursuing sustainable wellbeing: that is, wellbeing for all current peoples as well as for future generations. This means paying attention to its distribution between peoples, and to issues of equity and social justice. Between an upper boundary set by biophysical limits and a lower boundary set by decent levels of wellbeing for all today lies a safe and just space for humanity. The chapter concludes by noting two global landmarks in 2015: the UN Sustainable Development Goals (SDGs) and the Paris climate agreement. Together they reveal a yawning gap between what is needed for a safe climate and the prospects for a just and flourishing society.

Chapter 2 sets out a novel normative framework to judge progress in both human welfare and planetary sustainability: universal human needs and their satisfaction. It outlines a theory of human need and identifies health and autonomy as fundamental needs universally required to enable people to participate in their social forms of life. It goes on to distinguish these universal needs from culturally specific satisfiers and sketches a way of assessing the latter. These material satisfiers themselves require a set of institutions to decide on and produce them in a sustainable way. Finally, the chapter restates the strong normative priorities that follow: meeting people's basic needs, now and in the future, should be the first priority of justice, and satisfying needs thus takes moral precedence over satisfying consumer preferences.

Chapter 3 develops a political economy approach to understand 'climate capitalism', a model that aims to square capitalism's need for profit and continual growth with rapid decarbonisation of the world economy.

It analyses the major global drivers of emissions, including population growth, income growth, the eco-efficiency of production, and the global divide between emissions from production and consumption. It then turns to the role of inequalities – international and intra-national – and their impact on emissions and responsibilities for global warming. It outlines and critiques the current dominant perspective of 'green growth' powered by investment in renewables and carbon-saving technological change designed to decouple emissions from output. The chapter concludes by noting the current three-way contradiction between economic growth, ending poverty and dangerous climate change.

Chapter 4 discusses some of the questions, dilemmas and opportunities that arise when the claims of human need confront the present global economic system. It asks what would constitute a moral minimum of need satisfaction across today's world and then tries to estimate what 'necessary emissions' that would entail. Meeting needs will always be a lower carbon path than meeting untrammelled consumer preferences financed by ever-growing incomes. But whether it is low enough to protect the needs of future generations will depend, first, on the conflicts and synergies between the SDGs and a 2 °C mitigation strategy and, second, on the presence of a global equity framework. All existing strategies ignore the role of consumption in the affluent world, yet sustainability and distribution are intimately connected. My conclusion is that equity, redistribution and prioritising human needs, far from being diversions from the basic task of decarbonising the economy, are critical climate policies.

Chapter 5 starts with social policies and their embodiment in Western 'welfare states'. What are the new climate-related risks we can expect in the developed world and what are their implications for social policy? The chapter traces the development of welfare states and shows how they are being eroded by external and internal pressures, and have been outflanked by a rise in inequality. It applies comparative policy analysis to outline parallels between 'climate mitigation states' and welfare states. Such a survey reveals both common trends and significant national and regional variations. The chapter concludes by distinguishing three routes to decarbonisation – green growth, recomposed consumption, and degrowth. It sets up a framework for tracking the relationship between climate policy and social policy within these routes, which is applied in the remaining chapters.

Chapter 6 surveys climate mitigation programmes to reduce territorial emissions in the global North, building on the discussion of green growth in Chapter 3. It describes current policy frameworks for cutting carbon and surveys the major carbon mitigation strategies: pricing carbon, regulation, and strategic investment. It then charts some of the distributive and social consequences of these policies and the roles that social policies can and cannot play in counteracting them. It calls for a move from reactive social policies to integrated 'eco-social' policies, such as 'green new deals' to retrofit housing and provide sustainable domestic energy. It concludes that radical and fair carbon mitigation will require a shift from the neoliberal model towards a more coordinated and actively interventionist state.

Chapter 7 turns from production to consumption and consumption-based emissions. This leads to another policy goal for the rich world: to 'recompose' consumption to make it more sustainable. Yet simply redistributing income to low-income households could raise, rather than lower, emissions. This chapter therefore returns to the theory of human need. It sets out a 'dual strategy' methodology for identifying a *minimum* bundle of necessary consumption items in the UK and suggests how it might be used to identify a *maximum* bundle for sustainable consumption. In this way a 'consumption corridor' between upper unsustainable and lower unacceptable bounds can be charted. In the light of powerful corporate and other interests shaping consumer preferences a broad strategy of upstream prevention is advocated. To implement this approach, further eco-social policies are suggested, including taxing high-carbon luxuries, more social consumption and household carbon rationing. The conclusion notes that this whole approach challenges some fundamental principles of orthodox economics.

Chapter 8 takes off from the argument that even a combination of radical eco-efficiency and recomposing consumption will not cut emissions fast enough to avoid dangerous global warming if economic growth continues in the rich world. It sets out some basic features of a post-growth or degrowth economy: an emphasis on reproduction not production, investment not consumption, more discretionary time not more commodities, more equality and redistribution not less. This would profoundly affect all existing welfare states, which are premised on economic growth. A variety of policy solutions are considered, including spreading wealth more evenly through alternative forms of taxation and ownership, and fostering the core or social economy. An economy and a society that can no longer rely on annual growth will require a radical redistribution of carbon, time and wealth. The most realistic policy to achieve this transition, it is argued, is gradually to reduce paid work time – another vital eco-social policy.

Chapter 9 concludes. The idea of common human needs provides an essential alternative to the pursuit of unsustainable consumption growth within contemporary capitalism. Needs are limited; wants are limitless. Yet the pursuit of social welfare and climate stability today cannot be separated from the dynamics and future of capitalist economies. The chapter advocates a three-stage process to reconcile human wellbeing with planetary stability. The first, more eco-efficient green growth, requires a shift from liberal to more coordinated forms of capitalism. The second, recomposing consumption, would require at the least a shift from coordinated to a more 'reflexive' form of capitalism. The third, degrowth, is incompatible with the accumulation drive of any form of capitalism yet is ultimately – and quite soon – essential for our future prosperity, if not our very existence. It is for this reason, among others, that this book proposes an interim strategy to recompose consumption in rich countries towards low-carbon need satisfiers. It could provide a viable route from a dangerous present to a seemingly impossible future.

PART I

Concepts and global issues

1. The social dimensions of climate change

CLIMATE CHANGE AND HUMAN WELLBEING

Climate change threatens human wellbeing across the world and into the future. It poses an existential challenge with no past parallels: 'a truly complex and diabolical policy problem' (Steffen 2011). Human wellbeing obviously depends on Earth's support systems, and these are many and varied and, crucially, interactive. For the past 10 000 years – the Holocene – these systems have maintained a relatively stable state, forming the ecological foundations for the emergence of human civilisation. Now a new geological epoch, the Anthropocene, has been identified, where human activities start to have a significant – and negative – global impact on Earth's ecosystems.

The Stockholm Resilience Centre has identified nine critical Earth-system processes: climate change, the rate of biodiversity loss, the nitrogen and phosphorus cycles, stratospheric ozone, ocean acidity, global freshwater supplies, agricultural land availability, atmospheric aerosol loading, and chemical pollution. It goes on to estimate safe boundaries for each and concludes that the first three of these 'planetary boundaries' have already been crossed. Climate change – the change in energy in the atmosphere due to greenhouse gas emissions (the degree of 'radiative forcing') – already exceeds safe boundaries (Rockström et al. 2009). As I finish this chapter, two news reports have announced that 2016 will be the warmest year ever recorded, following on from the previous warmest, 2015, and that temperatures over the Arctic are an astonishing 20 °C higher than normal.

Of these limits, this book is concerned only with climate change. Choosing just one of the nine ecological boundaries is not without problems: it can sideline major socio-ecological issues such as food, water and the state of the oceans. The choice is partly due to lack of time, space and knowledge. But it also reflects a consensus that climate change is the great 'threat multiplier', posing the most immediate, serious and intractable threat to human wellbeing in today's world. It is this relationship between human wellbeing and climate change that is the primary focus of the book.

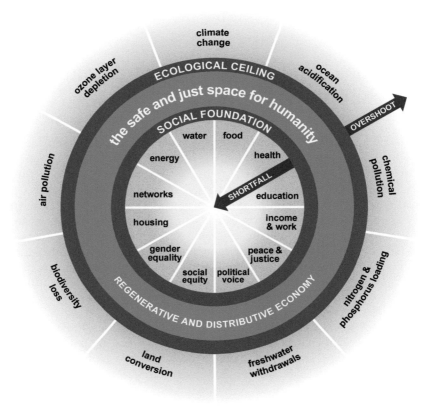

Source: Raworth (2017: 44).

*Figure 1.1   A safe and just space for humanity to thrive in: a first
illustration*

Kate Raworth pictures the interaction of planetary boundaries and human
wellbeing as a 'doughnut' or 'lifebelt', illustrated in Figure 1.1.

The nine planetary boundaries described by the Stockholm Resilience
Centre constitute the outer circle (Rockström et al. 2009). The social
foundations of human wellbeing constitute the inner boundary, drawing
on the 'Sustainable Development Goals' formally accepted by the United
Nations General Assembly in 2015. The exact nature of these components
of human welfare will be discussed below and in Chapter 2.

The most urgent global task is to bring everyone above the social
foundations which guard against threatening social deprivations while

not exceeding the critical planetary boundaries – which will in turn guard against future generations falling below these social foundations.

> Between a social foundation that protects against critical human deprivations, and an environmental ceiling that avoids critical natural thresholds, lies a safe and just space for humanity – shaped like a doughnut (or, if you prefer, a tyre, a bagel, or a life saver). This is the space where both human well-being and planetary well-being are assured, and their interdependence is respected . . . Moving into the safe and just space for humanity means eradicating poverty to bring everyone above the social foundation, and reducing global resource use, to bring it back within planetary boundaries. Social justice demands that this double objective be achieved through far greater global equity in the use of natural resources, with the greatest reductions coming from the world's richest consumers. And it demands far greater efficiency in transforming natural resources to meet human needs. (Raworth 2012: 5, 12)

Yet there is profound disjuncture between the ecological and social imperatives. The planetary boundaries remained unviolated in the Holocene for about 10000 years until recently. But there have always been, and there remain, millions of people in dire need, hungry, sick, impoverished, threatened, insecure, harmed in numerous objective ways and suffering in innumerable personal and collective ways. It is quite conceivable to pursue programmes for planetary sustainability without tackling this backlog of unmet needs; indeed there are many ways in which climate policies may harm the poor and worsen global inequality. For the moment, I simply assert that there are profound moral and consequential arguments for linking the two together and tackling both at once.

This book places the social dimension of wellbeing firmly in the centre of discussions about climate change – and vice versa. This chapter sets the scene by first summarising our best knowledge about the past, present and predicted future(s) of global warming. Second, it sketches the likely impacts of future climate change on human habitats and human welfare. Third, it summarises the options for climate policy, in terms of mitigating climate change and adapting to it. Fourth, it looks in more detail at the inner ring of Raworth's lifebelt and considers how far the Sustainable Development Goals agreed in 2015 provide a framework for conceiving and measuring these policy options.

THE CHALLENGES OF CLIMATE CHANGE

The Intergovernmental Panel on Climate Change (IPCC) brings together the majority of climate scientists in the world and issues reports of the global scientific consensus, the latest being the Fifth Report, published

in 2013/14. It marks an unprecedented mobilisation of scientific cooperation, and provides a benchmark of current best knowledge on which to
draw (see also Royal Society 2010). The arguments of climate deniers, still
powerful in a few countries (mainly in the 'Anglosphere', and notably the
US), are not considered here. However, their role is discussed in places as
a feature of political and ideological efforts to block and delay attempts to
curb climate change.

In summary, mean global temperatures are now about 1.0 °C higher
than nineteenth-century levels. The 2014 IPCC Summary Report states:

> Warming of the climate system is unequivocal, and since the 1950s many of the
> observable changes are unprecedented over decades to millennia. The atmos-
> phere and oceans have warmed, the amounts of snow and ice have diminished
> and sea level has risen ... Atmospheric concentrations of carbon dioxide
> (CO_2), methane and nitrous oxide are unprecedented in at least the last 800,000
> years. (IPCC 2014a: 1, 4)

The scientific link between this warming and certain atmospheric
emissions – labelled 'greenhouse gases' (GHGs) – has been established
for over a century. The most significant is carbon dioxide or CO_2, which
accounts for three-quarters of the total and is growing. CO_2 is mainly
the product of burning fossil fuels (almost 80 per cent) and of certain
industrial processes such as cement production, refineries, the iron and
steel industry and the petrochemical industry. The remaining quarter
of anthropogenic GHG emissions are non-CO_2 gases, notably methane
and nitrous oxide stemming from agriculture and other human uses of
land.

Future global warming will depend on the cumulative stock of green-
houses gases in the atmosphere. All GHGs stay in the atmosphere and
thus contribute to global warming for some time, but the time period
differs. Methane, though a potent GHG, lasts only about 12 years, but
'carbon is forever'. A total of 65–80 per cent of CO_2 released into the air
dissolves slowly into the ocean over a period of 20–200 years; the remainder can take up to several hundreds or thousands of years to disappear.
Given this fact, it is the total carbon budget – the cumulative total of
emissions – that matters, not the target for some year in the future, such as
2050. The IPCC (2014a: 8) concludes: 'Multiple lines of evidence indicate
a strong, consistent, almost linear relationship between cumulative CO_2
emissions and projected global temperature change to the year 2100.' In
the absence of effective climate action the major driver of this climatic shift
will be cumulative emissions of CO_2, which will 'largely determine global
mean surface warming by the late 21st century and beyond'.

Current predictions by the IPCC (2013: 10) are pessimistic:

Surface temperature is projected to rise over the 21st century under *all* assessed emission scenarios. It is 'very likely' that heat waves will occur more often and last longer, and that extreme precipitation events will become more intense and frequent in many regions. The ocean will continue to warm and acidify, and global mean sea level to rise. (IPCC 2013: 10)

Moreover, most aspects of climate change will persist for many centuries *even if* emissions of CO_2 are stopped. Past, present and future emissions of CO_2 commit us to mitigation action for centuries to come.

HUMAN IMPACTS

The IPCC goes on to analyse the likely impacts of global warming on a) ecosystems and b) human welfare, using integrated assessment models (IAMs). Impacts clearly attributable to climate change, as opposed to other causes, include: permafrost warming and thawing in high-latitude and high-elevation regions; shifts in the geographical ranges of many terrestrial, freshwater and marine species; negative impacts on crop yields (exceeding positive impacts); and impacts of recent climate-related extremes – such as heatwaves, droughts, floods, cyclones and wildfires – on many human systems (IPCC 2014b: 4–6).

The uncertainties involved in such modelling should always be borne in mind. The wide range in predicted temperatures has been noted, but the uncertainties multiply when forecasting things like crop yield changes and impacts on biodiversity, and still more when estimating human impacts and the effects of climate-related strategies. The predictions depend on at least three estimations: first, the extent to which impacts can be *attributed* to future climate change; second, assumptions about the impacts of current and future *mitigation* strategies; and, third, assumptions about the impacts of current and future *adaptation* strategies (IPCC 2014b: 8).

This global modelling is the starting point for most studies of the impacts of climate change on human habitats and human welfare. Another useful approach focuses on social vulnerability, starting from the local development context in which climate change occurs. The social and health impacts of climatic hazards are always mediated by the vulnerabilities of populations and their capacities to 'respond, recover and prepare' (O'Neill 2016). This will depend on a wide range of factors, including personal factors such as age or health, community factors such as the intensity and support of social networks, and societal-level factors such as the distribution of wealth and power, the strength of solidarity and the robustness of collective institutions. If we are to capture the new social risks and threats to wellbeing resulting from climate change we need to combine both approaches (IPCC 2014c).

The IPCC (2014b) Second Working Group Report on Human Impacts covers a wide range of issues: freshwater resources, terrestrial and inland water systems, coastal systems and low-lying areas, ocean systems, food security and food production systems, urban areas, rural areas, key economic sectors and services, human health, human security, livelihoods and poverty. I will briefly consider just a few of these.

It is projected that global *food security* will be undermined by climate change. Wheat, rice and maize production in temperate regions will be negatively affected, though some higher latitudes will benefit from a longer crop season. Water resources will decrease in most dry subtropical regions. Sustainable fisheries will be threatened by a redistribution of marine species and a reduction in biodiversity. Tropical regions will be disadvantaged by the redistribution of potential fisheries towards higher latitudes.

Health will be affected directly by changes in temperatures and weather systems, indirectly by changing disease vectors and crop patterns, and also by social responses to these shifts, such as population displacement. Impacts include a greater likelihood of disease and death due to heatwaves and fires, as well as from food- and water-borne diseases. The increased threats to food security may result in more under-nutrition, especially in developing countries with low incomes. Vulnerable populations may also face reductions in work capacity and labour productivity, which can in turn affect health.

In terms of *livelihoods and poverty* climate change will act as a 'threat multiplier': poor, marginal and socially excluded groups will suffer more, with the deepest impacts likely to be felt in sub-Saharan Africa and South Asia. Higher food prices are particularly likely to affect households depending on wage labour that are net buyers of food, especially in urban areas. Extreme weather events, as well as flooding due to rising sea levels, will threaten the territorial integrity of small islands and states with extensive low coastlines. There is considerable historical evidence suggesting that changes in climatic conditions have already been a contributory factor in migration. This includes large population displacements in the wake of severe events such as the northern Ethiopian famines of the 1980s, Hurricane Mitch in Central America in 1998 and Hurricane Katrina in New Orleans in 2005. On the other hand vulnerability is inversely correlated with mobility: those most exposed and vulnerable to the impacts of climate change have the least capability to migrate. The resulting 'trapped populations' may well suffer more than migrants (Foresight 2011; IPCC 2014b). Other results, such as conflicts over scarce water resources, have the potential to increase rivalry between states.

People, assets and ecosystems in *urban* areas are likely to suffer the consequences of heat stress, extreme precipitation, flooding, landslides, air pollution, droughts and water scarcity. Exposed areas and populations living in poor housing conditions will once again be most at risk. *Rural* areas will also be affected by these weather events, as well as by changes in agricultural incomes and access to water. Those communities with limited access to land and to modern agricultural technologies and infrastructure are expected to be most negatively affected (IPCC 2014b).

In a previous report the IPCC (2007) concluded that climate change will impact most heavily on tropical and subtropical regions, where standards of living are in general lower. The Fifth Report is less willing than previous reports to draw together these risks into an overall global pattern of climate vulnerability. However, it does repeat that risks are 'generally greater for disadvantaged people and communities in countries at all levels of development'. The following 'hotspots' are identified: the Mediterranean Basin, Central America, Central and West Africa, northern high-latitude regions, the Amazon, south/west US, South-East Asia, and the Tibetan Plateau (IPCC 2014b). And the report highlights two 'unique and threatened systems' subject to very high risks with additional warming of 2 °C: the Arctic sea ice and coral-reef systems (IPCC 2014b: 12).

In conclusion, the IPCC Report is unambiguous: 'Continued emission of greenhouse gases will cause further warming and long-lasting changes in all components of the climate system, increasing the likelihood of *severe, pervasive and irreversible impacts* for people and ecosystems. Limiting climate change would require substantial and sustained reductions in greenhouse gas emissions' (IPCC 2014a: 8, italics added).

If current mitigation policies prove inadequate and we head for a 4 °C warmer world, then the prospects for humanity are dire according to the World Bank (2012): 'If the global community fails to act on climate change it will trigger a cascade of cataclysmic changes that include extreme heat-waves, declining global food stocks and a sea-level rise affecting hundreds of millions of people.'

Kevin Anderson (2012) warns that 4 °C is 'incompatible with any reasonable characterisation of an organized, equitable and civilized global community'. And Lord Stern, Head of the 2006 Stern Review on the Economics of Climate Change, writes:

Five degrees is absolutely enormous. It would redraw the physical geography of the world. Large parts of the world would become desert, including most of southern Europe and the southern part of France. Other areas would be inundated. You'd see massive movements of population . . . This isn't a black swan, a small probability of a big problem; this is a big probability of a huge problem. (In Kaul et al. 2009: 136)

It is not surprising that 'end times' and dystopian futures are a growing feature of fiction and other art forms in this new millennium (Hamilton 2010; Urry 2011).

Unequal Contributions to Climate Change

The sources and drivers of emissions are discussed further in Chapter 3. For the moment they can be summarised as follows.

First, the bulk of the *cumulative* stock of CO_2 in the atmosphere has been contributed by the rich industrialised world: the burning of fossil fuels has precisely been a major source of their wealth. Since 1850, over one-half of the global total has been emitted by the US and Europe. Because CO_2 remains in the atmosphere for many generations, cumulative emissions remain an important measure of responsibility for global warming.

However, second, in recent years the annual emissions of emerging market economies, most notably China, have expanded fast and over-taken those of the developed world, which have remained stable. China is now the world's largest annual emitter of CO_2.

But, third, there is a growing discrepancy between the emissions from a given national territory (production-based emissions) and the 'consumption footprint' of a given national territory (consumption-based emissions). Consumption-based emissions of the OECD are higher than their territorial emissions, while those of the rest of the world are lower. This reflects the outsourcing of manufacturing and industry from the West to the East during the period of intense globalisation – and accompanying deindustrialisation in the West – since around 1980. Consumption emissions of the West continued to rise – until the sharp fall occasioned by the severe financial crisis and recession of 2007–09. The West has gained doubly: both from rising consumption and from an apparently declining responsibility for the associated emissions. There has been 'environmental load displacement' from North to South (Christoff and Eckersley 2013).

Fourth, these country totals make little sense without taking into account their populations (though it is interesting how often this simple fact is ignored: one might as well praise tiny Luxembourg for its commendable emissions record when in fact it has the highest emissions per person in the world). When this is done, emissions per person reveal wide inequal-ities, especially when calculating consumption-based emissions: North Americans emit on average 22 tonnes of CO_2 a year, West Europeans 13 tonnes, Chinese 6 tonnes and South Asians and Africans 2 tonnes (Chancel and Piketty 2015). These gaps are closing, but at a slow pace.

Finally, predicted *future* annual emissions show parts of the developing world overtaking the developed. The national contributions to the accumulated stock of emissions will follow a similar pattern but with a lag.

Putting together national and regional responsibilities for global warming with its national and regional impacts reveals a global *double injustice* (Gough 2011a). Consumers in the richer countries of the West still account for about one-half of cumulative greenhouse gases and enjoy on average considerably higher standards of living than the rest of the world. Yet on balance they are predicted to suffer fewer negative climatic impacts over the next two decades. Indeed some areas, such as Northern Europe, could benefit from longer growing seasons and lower heating costs. Tropical and sub-tropical areas with lower incomes and few past responsibilities for emissions will suffer major negative impacts.

The globalisation of the world economy that has brought this about suggests that the old duopoly of 'global North' and 'global South' is no longer appropriate when discussing global climate patterns. To reflect this, in the rest of the book I will use instead a threefold distinction between *North*, *East* and *South*:

- The *North* comprises the original OECD member states and some developed newer members (e.g. South Korea).
- The *'East'* comprises the fast-emerging economies including the BRIC countries (Brazil, Russia, India, China) and the MINT countries (Mexico, Indonesia, Nigeria, Turkey).
- The *South* comprises the developing world of low-income and low-middle-income countries, mainly in Africa, Asia and parts of Latin America.

These terms make no geographical sense – of course Australia is not in the north, Brazil and Mexico are not in the east – but all nomenclatures have problems and these have the virtue of brevity.

CLIMATE POLICY

Standing between this powerful dynamic of growing emissions and the dangerous consequences of global warming are a range of 'climate policies', at global, regional, national, sub-national and sectoral levels. Figure 1.2 offers a very simplified portrayal of the complex links between the causes of growing emissions and the impacts on human wellbeing. This helps to distinguish three categories of climate policies.

Climate *mitigation* policy refers to human interventions to reduce the sources, or enhance the 'sinks', of greenhouse gases. Broadly speaking,

Economic activity ->
  Energy consumption ->
    Greenhouse gas emissions -> -> Absorption by GHG sinks
      GHG cumulative concentrations ->
        Global temperature rise ->
          Regional and local climate change ->
            Impact on human habitats ->
              Human wellbeing

Figure 1.2   From human activity to climate change to human wellbeing

mitigation policies target the first three rows of Figure 1.2: to reduce emissions and enhance global carbon sinks, requiring shifts in energy policy and, further upstream, the scale and nature of economic activity. The language of mitigation is used rather than 'prevention', in recognition of the fact that further global warming cannot be prevented owing to the accumulation of past GHGs in the atmosphere.

Climate *adaptation* policy seeks to lower the risks posed by the consequences of climatic changes. Adaptation policies target the last two rows of the figure to reduce risks to habitats and human wellbeing. Clearly a combination of mitigation and adaptation interventions will be necessary to forestall harmful impacts on human populations.

In between the two there is a potential third domain of interventions designed to directly target cumulative GHG concentrations, global temperature rise and regional climate change. This is the arena of *geo-engineering*, deliberate large-scale interventions in the Earth's natural systems to counteract climate change, for example by reflecting part of the sun's energy back into space or directly removing CO_2 from the atmosphere. Geo-engineering is beginning to enter policy debate, but the current consensus is that it is fraught with unforeseeable and potentially catastrophic consequences and I do not discuss it further here. 'We should remain under no illusion that if we have to resort to these kinds of technologies, then humankind is in a mess' (Berners-Lee and Clark 2013; cf. Royal Society 2010).

Mitigation

Mitigation plays a central role in Article 2 of the United Nations Framework Convention on Climate Change (UNFCCC), which aims for 'stabilisation of greenhouse gas concentrations in the atmosphere at a level that would prevent dangerous anthropogenic interference with the climate system'. The Copenhagen Accord was endorsed by 167 countries, agreeing that the safest maximum amount that global temperatures should be allowed to rise above the pre-industrial level is 2 °C. To achieve this, the Paris agreement in December 2015 calls for zero net anthropogenic GHG emissions to be reached during the 'second half of the 21st century'. In addition nations at Paris agreed to drive efforts to limit the temperature increase to no more than 1.5 °C above pre-industrial levels (UNFCCC 2015).

This is a safer but dauntingly ambitious goal. According to Lord Stern (2015), to achieve (only) a 50:50 chance of avoiding global warming exceeding 2 °C by the end of the century, and taking population growth into account, global emissions must be cut from around 7 tonnes CO_2e[1] per person per year now to no more than 2 tonnes by 2050 – a revolutionary downshift of 3.5 times. But, if output per person continues to grow at its present rate (roughly trebling by 2050), then global emissions *per unit of output* must fall by a factor of 7–8 times by 2050 – as I write now only 33 years away.

Moreover, the risks of modelling a 50:50 chance are self-evident. Bill McKibben (2012) has taken to task other models using a 'reasonable 20%' risk, pointing out that 'reasonable' in this case means 'one chance in five, or somewhat worse odds than playing Russian roulette with a six-shooter'. A 50:50 chance is like playing Russian roulette with bullets in three chambers.

The range of potential mitigation interventions is enormous. A short list would include: reducing fossil fuel extraction; pricing carbon; fostering renewable energy technology, alternative fuels and alternative transport systems and mobility structures; investing in new technology and energy efficiency; encouraging net forestation, changes in rural land use and agricultural practices (management of croplands, grazing lands and soil restoration); managing urban forms and land use; building infrastructure and spatial planning; designing buildings for energy efficiency; and changing consumer behaviour, lifestyle and culture (IPCC 2014c). Some of these will be encountered and analysed in this book.

Taking just the first of the list above, research by Carbon Tracker estimates that the world could emit about 900 gigatonnes (Gt) (billions of tonnes) of CO_2 between 2000 and 2050 (of which 17 years have now elapsed) and still have a 'reasonable' chance of avoiding 2 °C warming (Ranger and Ward 2013). These figures are far lower than present-day estimates of the carbon embedded in usable reserves – some 2860 Gt, implying that two-thirds of present reserves of coal, oil and gas cannot be mined. This has fostered a new approach to carbon mitigation: to 'keep the oil in the soil' (Berners-Lee and Clark 2013).

---

[1] CO_2e stands for CO_2 equivalent, or the amount of a given gas that would have the same warming potential as a tonne of CO_2.

Adaptation

Adaptation is defined by the IPCC as 'the process of adjustment to actual or expected climate and its effects' in both human and natural systems (IPCC 2014c). This process of adjustment differs according to regions and includes: the evolution of agricultural practices adapted to temperature; agroforestry to manage wildfires; land-use changes and settlement relocation; increased efficiency in water management systems; infectious disease control; wetland restoration and the maintenance of coastal landforms; enhanced monitoring, regulation and early warning systems for extreme weather events; stronger flood defences; the development of sustainable cities; and so on. This list excludes social adaptation measures such as building social capital and more resourceful communities able to withstand climate impacts.

An important distinction between the two policy domains must be noted here (Kolstad et al. 2014): mitigating GHG emissions is predominantly a 'global public good', whereas adaptation is more often a private or national good. Gains from adaptation (such as changing a crop to one more heat tolerant, or building flood defences) tend to be realised by the same parties or territories that are incurring the costs. There may be externalities involved, but these tend to be more localised and contemporaneous.[2]

The case of GHG mitigation is quite different: emissions in any geographical space will affect the global concentration everywhere. There are thus 'collective action problems'. 'Incentives for individuals or countries to unilaterally reduce emissions are considerably reduced; free-riding on the actions of others is a dominant strategy . . . and lack of coordination yields insufficient mitigation' (IPCC 2014a). Elinor Ostrom's (1990) research on common pool resources concludes that efficient environmental management is more likely where four conditions hold: the environmental problem is visible; cause and effect relations are understood; the problem is reversible; and management results in clear net benefits to key constituencies. For many decades after the scientific community began to chart global warming none of these conditions applied to mitigating climate change, outside a few pioneers. Awareness of the first two is now growing as conditions worsen (Christoff and Eckersley 2013).

---

[2]   This of course by no means removes the presence of interests, power and realpolitik from the political economy of adaptation (Sovacool and Linnér 2016).

SOCIAL CONDITIONS AND HUMAN DEVELOPMENT

Turning to the 'inner ring' of Raworth's (2012) lifebelt, I now summarise international and global efforts to map out the 'social' domain. What 'welfare' or 'wellbeing' consists of is discussed in detail in Chapter 2, but they certainly include concerns with health and survival chances, literacy and learning, access to essential resources and opportunities to participate in social life. The social dimension typically combines two things: a concern with *levels* of human wellbeing, and a concern with equity and justice – the way that wellbeing is *distributed* between peoples.

There is clear evidence of improved global wellbeing since the Second World War, as the 2015 Rockefeller Foundation–Lancet Report makes clear (Whitmee et al. 2015):

> By most metrics, human health is better today than at any time in history. Life expectancy has soared from 47 years in 1950–1955, to 69 years in 2005–2010 . . . The total number of people living in extreme poverty has fallen by 0.7 billion over the past 30 years, despite an increase in the total population of poor countries of about 2 billion. This escape from poverty has been accompanied by unparalleled advances in public health, health care, education, human rights legislation, and technological development that have brought great benefits, albeit inequitably, to humanity.

The United Nations Development Programme (UNDP) has developed a composite measure, the Human Development Index (HDI), that combines measures of health (life expectancy), learning (literacy and education) and resources to meet other basic needs (the log of GDP per head). This also charts significant improvements across the world, though less so in Africa.

Against this picture of progress, income inequality between countries widened throughout most of the twentieth century as the industrialising West pulled away from the rest of the world. It had also colonised much of that world or exercised economic power to prevent the emergence of serious competition. Only around the turn of the millennium did this trend reverse, when economic growth in the East really took off (Bocchiola 2013). However, since around 1980 inequality has begun to rise *within* most countries, in some at a rapid rate. This has renewed concerns over equity and justice and slowed down the rate of global average improvement in the HDI – by one-quarter from what it would otherwise be, according to the UNDP (2011).

Global concern with social conditions has taken many forms, including efforts to establish global social rights and agreed development goals. The 1948 UN Declaration on Human Rights comprised a wide range of civil and political rights but also certain economic and social rights: for example the rights to work, education and a standard of living adequate for health and wellbeing, including food, clothing, housing, medical care, necessary social services, and the right to security in the event of unemployment, sickness, disability, widowhood, old age or other lack of livelihood in circumstances beyond people's control (UN General Assembly 1948). This was followed by a further stream of declarations and conventions, many codified in the International Covenants on Civil and Political Rights (ICCPR) and Economic, Social and Cultural Rights (ICESR), both of which came into force in 1976. Since then, there have been the 1981 International Convention on the Elimination of All Forms of Discrimination against Women (ICEDAW), the 1989 Convention on the Rights of the Child and others. All of these include social and economic rights. However, implementation is soft, to say the least.

In 2000 the UN Millennium Development Goals (MDGs) were unanimously agreed by all 189 member states (Hulme and Scott 2010). Eight goals were set with 21 targets – for example to reduce by two-thirds by 2015 the under-five mortality rate; to halve, by 2015, the proportion of people without sustainable access to safe drinking water and basic sanitation – and more than 60 indicators were agreed to chart progress. They amount to 'a minimum set of global social standards in education, health and poverty alleviation' (Deacon 2007: 173). The poverty target has been achieved (ten years early) but not that for malnutrition. The gender equality goal was achieved and the education goal narrowly missed. However, the health goals were not met.

TOWARDS SUSTAINABLE DEVELOPMENT?

The global picture painted so far charts three diverging trends: first, significant progress in some core features of social conditions and human development across the globe; second, rising inequalities; and, third, an as yet uncontrolled accumulation of greenhouse gases that will drive future global warming with significant damaging consequences for habitats and human welfare.

The following questions arise:

- What are the implications of climate change for future human welfare?
- How will the pursuit of climate stability interact with the pursuit of social improvement?
- Can we achieve some combination of equity/justice and sustainability?

These questions lie at the heart of this book.

It is interesting that two reports approaching the question from opposite directions come to roughly the same conclusion. The 2005 Millennium Ecosystem Assessment (Corvalan et al. 2005) concluded that, over the past 50 years, humans have changed ecosystems more rapidly and extensively than in any comparable period of time in human history, largely to meet rapidly growing demands for food, fresh water, timber, fibre, ores and fuel. This has resulted in a substantial and largely irreversible loss in the diversity of life on Earth. There have been substantial net gains in human wellbeing and economic development, but these gains have been achieved at growing costs in terms of the degradation of many 'ecosystem services', increased risks of non-linear changes, and the exacerbation of poverty for some groups of people. These problems, unless addressed, seriously threaten future human wellbeing.

The 2015 Rockefeller Foundation–Lancet Commission, writing about health and climate change, summarises recent decades of the relationship between welfare and sustainability as follows:

> We have been mortgaging the health of future generations to realize economic and development gains in the present. By unsustainably exploiting nature's resources, human civilisation has flourished but now risks substantial health effects from the degradation of nature's life support systems in the future . . . In essence, humanity has traded off many of the Earth's supportive and regulating processes to feed and fuel human population growth and development.

Looking ahead, the 2011 UNDP Report presents two simulations of the effect of unmitigated climate change on human development as measured by the HDI (not GDP) (see Figure 1.3). The first, the 'environmental challenge' scenario, models the adverse effects of global warming on agricultural production, on pollution and on access to clean water and improved sanitation. It suggests that by 2050 the global HDI would be 8 per cent lower than in the baseline case (and 12 per cent lower in South Asia and sub-Saharan Africa). A second, even more adverse, 'environmental disaster' scenario envisages extensive deforestation and land degradation, dramatic declines in biodiversity and accelerated extreme weather events. Here, global HDI would be some 15 per cent below the baseline. It would lead to a turning point before 2050 in poorer developing countries: their slow convergence with rich countries in HDI achievements would turn into an absolute decline.

Global policy responses to this twin threat can be dated back to the 1987 Brundtland Report on sustainable development and to the 1992 UN Rio Earth Summit. The former's famous definition of sustainable development – headlined in Chapter 2 below – provides an optimistic

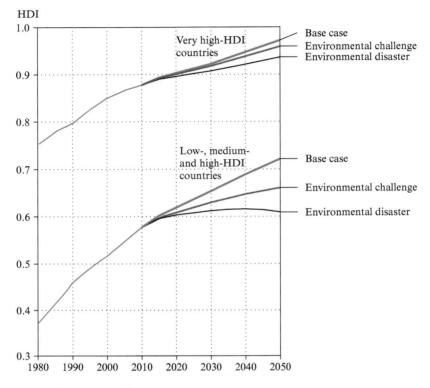

Source: UNDP (2011: 31).

Figure 1.3   Scenarios projecting impacts of environmental risks on human development to 2050

vision of a common goal for all humanity and forms the basis of a core argument in this book. The UN Rio Earth Summit in 1992 followed it up with a quite radical statement of normative principles to guide global environmental and climate change policy. These included the precautionary principle (Principle 15), the 'polluter pays' principle (Principle 16) and Principle 7 on 'common but differentiated responsibilities' between developed and developing countries (Christoff and Eckersley 2013).

But realising all this in practice since the Rio Summit has proved extremely difficult. The discourse of sustainable development has been interpreted as a political construct designed to facilitate a bargain across the deep structural divide between North and South (Vogler 2007). In simplistic terms, the North did not want to recognise the role of high incomes and consumption in climate change, while the South wanted to avoid association with population growth. What was left and could be agreed upon was, as always, technology and the role of human ingenuity in extracting more and more growth with fewer and fewer emissions. Twenty years later at the Rio+20 conference, sustainable development had been displaced by the weaker bridging concept of 'green growth': a strategy to decarbonise today's economies but within a capitalist framework committed to economic growth. The change from 'sustainable development' of Rio 1992 marked a shift from an environmental to a more economic discourse (see Chapter 3).

2015: New York and Paris

Two notable global agreements occurred in 2015: the unanimous adoption at the UN in New York of the Sustainable Development Goals (SDGs) and the Paris agreement in December on greenhouse gas emissions within the UNFCCC. Does this mark a new turn towards a sustainable pathway for people and planet?

The ambition of the SDGs has been described as 'breath-taking': a list of 17 goals and 169 targets, it is arguably the most comprehensive global agenda adopted since the UN Charter in 1945. Its thematic repertoire ranges from poverty, health, education and inequality through to energy, infrastructure, climate change, marine resources, peace, security and good governance. The goals have been criticised by Charles Kenny of the Center for Global Development as unmanageable, 'overwrought and obese': monitoring progress will be all but impossible and will encourage picking and choosing between the 169 targets.

Yet the UN Secretary-General welcomed the agenda by saying it 'encompasses a universal, transformative and integrated agenda that heralds an historic turning point for our world' (Fukuda-Parr 2015; Langford 2016). Unlike the MDGs, many targets apply to the North as well as to the South. The agenda is also ambitious: for example, one goal is to abolish poverty by 2030, not to reduce it. The agreement of 193 countries may perhaps signal the emergence of a broader normative consensus. Above all, the SDGs begin to take the social conditions of *future* generations seriously. Global warming brings intergenerational welfare, equity and justice centre-stage. Failure to counteract it will threaten the wellbeing of future generations, albeit in varied and unequal ways.

The 2015 SDG agenda concurs with the 2014 IPCC Report in recognising that some mitigation efforts can and do undermine actions to pursue equity or eradicate poverty: a comprehensive assessment of climate policies means evaluating the sustainability and equity of development pathways more generally (IPCC 2014c: 5). As the Brundtland Report (WCED 1987) definition makes clear, it is not acceptable to provide for current welfare at the expense of future welfare. But nor is it acceptable to provide for future welfare at the expense of the current welfare of the poor of the world. The SDG agenda arguably begins the process of relating – if not reconciling – environmental sustainability with social justice.

In late 2015 the Paris climate conference (COP) achieved a treaty agreement on several key issues, of which two are worth noting here. First, all states have clearly accepted the ultimate goal of *net zero emissions*, to be achieved 'in the second half of this century'. Second, every five years a stocktake will take place showing how far current emissions projections diverge from a 2 °C – and a 1.5 °C – pathway. Two years after each stocktake, all countries will have to produce, for the subsequent 10- or 15-year period, new targets and plans that are 'more ambitious' than the last. Countries' commitments are not legally binding, but the requirement to improve national plans every five years is legally binding. This architecture is designed to send clear signals to business, investors and governments that the future direction of the global economy is low carbon.

The agreement was ratified in November 2016, by which time 55 nations accounting for at least 55 per cent of total global GHGs deposited their articles of ratification with the United Nations. Opinions about the Paris COP of December 2015 vary widely. According to many participants and commentators the outcome was impressive: 'the Agreement lies at almost the highest level of ambition that could possibly have been achieved' (Jacobs 2015).

For others the Paris agreement is too little, too late. The American environmental activist Bill McKibben, for example, commented:

> the world emerges, finally, with something like a climate accord, albeit unenforceable. If all parties keep their promises, the planet would warm by an estimated 3.5 degrees Celsius . . . and that is way, way too much . . . The irony is, an agreement like this adopted at the first climate conference in 1995 might have worked. (McKibben 2015)

Critical environmentalist Clive Spash (2016) is uncompromising: 'The Paris agreement changes nothing . . . it treats worst case scenarios as an acceptable 50:50 chance . . . the targets and promises of the Paris Agreement bear no relationship to biophysical or social and economic reality.'

Perhaps the environmental journalist George Monbiot (2015) sums this up best: 'By comparison to what it could have been, it's a miracle. By comparison to what it should have been, it's a disaster.'

CONCLUSION

Climate change cannot be the basic cause of poverty, ill-health, unmet basic needs and fragile livelihoods; these have existed throughout human history. But the hazards of uncontrolled climate change constitute an epochal 'threat multiplier'. It will make the pursuit of economic and social needs and rights more difficult and, if global warming exceeds a threshold around 2 °C, then it will overwhelm all attempts to eradicate poverty, let alone provide all peoples with an acceptable level of security and a flourishing life.

The goal must be to respect biophysical boundaries while at the same time pursuing sustainable wellbeing: that is, wellbeing for all current peoples and for future generations. An acceptable and sufficient *level* of human wellbeing also means paying attention to its *distribution* between peoples. Issues of equity and social justice are central. The social dimension of climate change thus refers to both the average level of human wellbeing in any specific dimension and its just distribution. Both, together with respecting biophysical boundaries, are necessary for 'sustainable wellbeing'.

In summary, the pursuit of wellbeing and social justice is inadequate if it is at the expense of the biosphere and future generations. At the same time, the pursuit of human wellbeing for some while also respecting planetary limits is unacceptable if it is at the expense of global justice and the poor of the world. Similarly, the pursuit of social justice within planetary limits is inadequate if justice is understood solely in procedural terms, such as greater civic rights and Western democracy, while ignoring more material aspects of wellbeing.

This book focuses on the interrelation between three goals: wellbeing and social justice within planetary boundaries. It considers what might be done to advance all three together.

2. Human needs and sustainable wellbeing

> Sustainable development is development that meets the needs of the present without compromising the ability of future generations to meet their own needs. It contains within it two key concepts: 1) the concept of 'needs' . . . 2) the idea of limitations. (Brundtland Report, WCED 1987: 43)

This chapter focuses on the inner circle of Raworth's lifebelt – the nature of human wellbeing and flourishing, and of deprivation and suffering. It proceeds in six stages. First, the meaning of 'wellbeing' is probed and current dominant approaches, including the orthodox economics theory of satisfying consumer 'preferences', are found wanting. Second, I outline a theory of universal human needs which enables us to evaluate and compare wellbeing across different global contexts and cultures and across (a reasonable number of) generations into the future. Third, while basic needs are universal, they are satisfied in countless different ways, which do vary across cultures and times. The chapter elaborates 'need satisfiers' and a collective methodology for identifying them in specific contexts.

Fourth, meeting human needs also requires a socio-economic system that produces and delivers the necessary and appropriate need satisfiers, distributes them in line with need – and ensures that all this does not threaten planetary limits. In lieu of discussing the full suite of procedural and material preconditions for need satisfaction, the chapter focuses on the meaning of 'sustainability' and the material satisfiers for sustainable wellbeing. Fifth, it considers how needs and societal preconditions can be mapped and measured by comparing them with the Sustainable Development Goals discussed in Chapter 1. Together, I claim, they can provide a more robust framework than Raworth's for her 'inner boundary' of human welfare, equity and justice.

Finally, I argue that needs – unlike consumer preferences, wants and happiness – are not morally neutral. They imply ethical obligations on individuals and claims of justice – universal rights and obligations – on social institutions. In the Anthropocene, they assert standards of *sufficiency* and the moral priority of human needs (present and future) over consumer wants or preferences.

UNDERSTANDING WELLBEING

The previous chapter argued that human wellbeing lies at the core of the social domain, but what do we mean by wellbeing? The older English term 'welfare' can be traced back to the fourteenth century, when it meant 'to journey well' and could indicate both happiness and prosperity (Williams 1983). Ideas that cluster around welfare or wellbeing can be traced further back to at least Aristotle and the Buddha. Today they stir up a hornet's nest of problems. Philosophers from distinct schools of thought, economists, psychologists, sociologists, poverty researchers, development studies specialists, theologians and many others have contributed to debates on the nature of human wellbeing, how to measure it and how to enhance it (see Alkire 2002; Gasper 2007; Gough et al. 2007; Boarini et al. 2014; Dover 2016).

Without a sound theory of what constitutes human wellbeing, we cannot begin to consider how to pursue or achieve it in the context of climate change. Here I briefly consider and critique four current perspectives on wellbeing.[1] These are:

- the satisfaction of consumer preferences;
- happiness;
- relational wellbeing;
- human capabilities.

I then go on to advocate a theory of human need.

Preference satisfaction theory, the dominant conception of wellbeing within market societies, rests on two normative foundations: that individuals are the best judges of their own preferences or wants, and that what is produced and consumed should be determined by the private consumption and work preferences of individuals. This appeals to intuitions of personal autonomy and freedom. However, it assumes inter alia that all individuals have 'given and complete preference functions' and that all seek to maximise their individual utility (Hodgson 2013).

As such this theory has been subject to numerous challenges. First, individuals are not necessarily the best judges of their wants if their knowledge or rationality is short of perfect. Second, our preferences change in light of the options we have available – the phenomenon of 'adaptive preferences'. Third, markets and other capitalist institutions themselves influence the evolution of values, tastes, preferences and even personalities, generating a

---

[1] To avoid too many references here, the reader is referred to Gough (2015a) and to further citations within it.

'circularity of evaluation'. Fourth, the fundamental assumption that every individual is actuated only by self-interest has been disproved. Fifth, specifying welfare entirely in terms of preferences flattens moral distinctions between the seriousness of choices. Finally, since the preferences of future generations cannot be revealed through their choices or behaviour, the theory is particularly unsuited to considering the wellbeing of future generations. My conclusion is that preference satisfaction cannot provide a logical, ethical or practical conception and measure of human wellbeing – and especially not when we must consider wellbeing on a global and intergenerational scale (Gough 2015a).

Happiness and subjective wellbeing. Subjective ideas of wellbeing include hedonic psychology, researches into life satisfaction, and the economics of happiness. This work has developed useful measures of subjective wellbeing and a mass of solid findings on its determinants, including the finding that, beyond a rather modest income level, growth in real incomes and consumption is associated weakly or not at all with happiness or subjective wellbeing (Easterlin 2001). But happiness theory and metrics face some similar problems to preference satisfaction theory. Adaptation is pervasive: the process of adjusting expectations to reality appears to be a universal feature of the human condition applying to both losses and gains and to individual and collective events. Moreover, there is evidence of cultural bias: national values of individualism are correlated with reported wellbeing, so that cultures evoking a 'modesty bias', as in some countries of East Asia, report lower wellbeing scores. Though these and other problems may be controlled for when comparing wellbeing within societies, they fatally undermine the ability of happiness to provide a measure of wellbeing across cultures and times.

Relational wellbeing. The perspectives of post-modernism, post-colonialism and post-structuralism envisage human wellbeing as discursively constructed, local and incomparable. They propound a different ontology: of people in groups collectively defining and creating their wellbeing in specific social contexts. Public reasoning and deliberation within groups, drawing on the local knowledge of peoples in a specific context, are crucial to understanding how social groups interpret and act upon common threats such as the effects of climate change. This approach is critical for an understanding of wellbeing in specific contexts and will be used in this book, but as a global ethic it is not generalisable. The moral consequence is cultural relativism – an acceptance of whatever goals a social group proclaims, irrespective of their ethical justifications or their implications for other groups, now or in the future. It cannot provide a universalisable ethic of wellbeing in the face of global inequalities and climate change (Doyal and Gough 1991; Gough et al. 2007; Chibber 2013).

The capability approach, first elaborated by Amartya Sen, conceives human wellbeing in terms of the substantive freedoms and opportunities that people possess. These 'capabilities' in turn rest on the *functionings* of people: 'an achievement of a person: what she or he manages to do or to be' (Sen 1985: 12). A person's capability then represents all the combinations of functionings that are feasible to that person – that she could choose. The larger the set of choices, the greater the level of wellbeing. The capability approach has mounted a powerful challenge to orthodox welfare economics, has helped to establish a more rounded conception of the human person than *homo economicus*, and has founded the only globally accepted alternative metric to GDP so far – the Human Development Index. In my view it marks a significant step forward.

Yet it suffers from a fundamental problem: it provides no means for identifying *basic* functionings or capabilities common to a group of people let alone to all people.[2] Sen famously rejects the search for, and lists of, universally valued functionings. Consequently the approach leaves scant protection for future generations while the unconstrained consumption of natural resources by current generations are wreaking havoc with the environment. Martha Nussbaum (2000) does argue for the universality of 'human functional capabilities' and is content to identify them in a cross-cultural way, but to justify this she relies ultimately on the language of 'need', without articulating what this means (Nussbaum 2006; Brock 2009). Her work is returned to below.

In the face of these problems, I contend that only a theory of universal human need can enable us to conceive, measure and compare human wellbeing across time and space. The definition of sustainable development in the Brundtland Report (WCED 1987) makes two things clear: first, that mitigating climate change must be confronted simultaneously with addressing global poverty and inequality; and, second, that the language of needs is central to the very question of sustainable development. Global warming undoubtedly explains part of the recent resurgence of interest in the idea of common human needs. Yet so far no fully convincing theory of human need has gained acceptance; indeed, many writers, including the authors of the Brundtland Report, refer to needs as though they were self-evident. They are not, and the next section presents a theory of human need, drawing on earlier work (Doyal and Gough 1991).

---

[2] Capabilities are also extremely difficult to operationalise: the 'capability set' of a person includes not only the 'opportunities to have and to be' that people actually choose but also the near-infinite counterfactual opportunities that were open to them that they did not choose.

A THEORY OF HUMAN NEED

In *A Theory of Human Need* (1991) Len Doyal and I identify a conceptual space of universal human need. We recognise cultural variety in meeting needs, but aim to avoid subordinating the identification of needs to such cultural contexts. Our essential premise is that all individuals, everywhere in the world, at all times present and future, have certain basic needs. These must be met in order for people to avoid harm, to participate in society and to reflect critically upon the conditions in which they find themselves. Only if we understand needs in this way – in universal terms, applied across time and place – can we plan for and measure progress towards our social and environmental goals, both globally and into the future. Figure 2.1 sets out the architecture of the theory, distinguishing five levels.

Basic Needs

The universality of need rests upon the belief that if needs are not satisfied then *serious harm* of some objective kind will result. This is not the same as subjective feelings like anxiety or unhappiness. It refers to functions not feelings. This harm implies obstacles to successful social *participation*. All our private and public goals are achieved on the basis of successful social interaction, past, present or future, with others. Participation in this very broad sense reflects the social character of human action. (In practice, participation will manifest itself in a very wide variety of ways and with hugely different levels of intensity.) It follows that participation in some form of social life without serious systematic limitations is our most basic human interest.

Basic needs are then the universal preconditions for effective participation in any form of social life. To do this a person must be able to formulate aims, understand how to achieve them, and act to strive to achieve them in practice. Whatever a person's goals, whatever the cultural practices and values within which she lives, she will require certain prerequisites or basic needs, in order to strive towards those goals.

In this way we identify, alongside social participation, *health* and *autonomy* as the most basic human needs. Survival is the most basic need, but all people require a modicum of physical and mental health for effective social participation. To complete a range of practical tasks in daily life requires manual, mental and emotional abilities, with which poor health usually interferes. Illness results in suffering one or more dimensions of disability – however different peoples label and explain their illnesses.

Basic autonomy can be defined as 'the ability to make informed choices about what should be done and how to go about doing it'. Whatever the

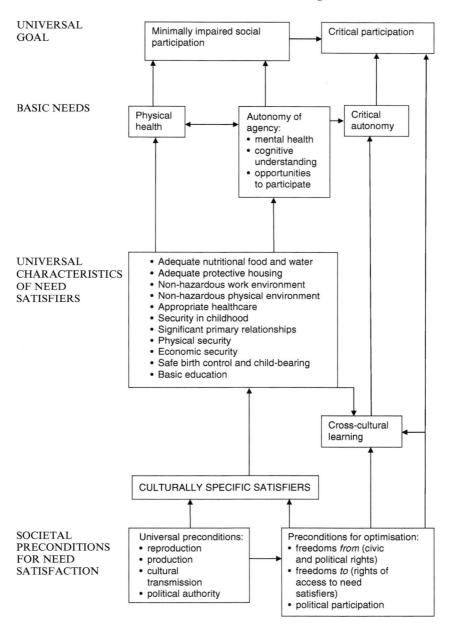

Figure 2.1 A theory of human need: universal goals, basic needs,
intermediate needs, satisfiers and societal preconditions

human contexts and predicaments, ranging from oppressive and totalitarian contexts to ones with wide options for creative participation, all can entail creative reflection. Indeed, poor and oppressed people living in shifting, uncertain and threatening environments must exhibit agency and creativity for much of their lives in order to achieve minimal goals.

Three key variables, we argue, affect levels of individual autonomy of agency. First, some level of *cognitive and emotional capacity* is a necessary prerequisite for a person to pursue a goal. This can be blocked by serious mental ill-health: the levels of rationality and responsibility present in the autonomous individual are undermined when a person suffers from severe mental illness. Second, an individual's autonomy is impaired if she lacks adequate *cultural understanding* of herself, her culture and what is expected of her as an individual within it. Learning is a universal process of human development, and we all need 'teachers' to develop, not just or even teachers in school but a range of significant others. Third, autonomy of agency requires a range of *opportunities* to undertake socially significant activities. By 'significant' we mean activities that are central in all societies. Braybrooke (1987) classifies these as the roles of parent, householder, worker and citizen. We restyle these four basic social activities as production, reproduction, cultural transmission and the exercise of political authority. To be excluded from participation in any of these domains is to have one's autonomy impaired.

In addition we distinguish a higher level of *critical autonomy*: the capacity to compare cultural rules, to reflect upon the rules of one's own culture, and to work with others to change them. At this higher level, drawing on imagination, past examples or comparisons with other ways of life, people can begin to question the taken-for-granted cultural frames of their own ways of life. This requires, beyond freedom of agency, some measure of political freedom.

Human need theory has been advanced from a variety of perspectives (Dover 2016). One of these is eudaimonic (as opposed to hedonic) psychology, which provides argument and evidence for three universal psychological needs: *autonomy*, *competence* and *relatedness*. Fulfilment of these needs is necessary for psychological 'wellness', and observable consequences follow from their lack of fulfilment (Ryan and Deci 2001; Ryan and Sapp 2007). For example, individuals whose life goals are more focused on wealth, image and fame than on relationships, personal growth and community evidence less self-esteem, self-actualisation and life satisfaction. More materialistic individuals are most likely to be dissatisfied with life, lack vitality, and suffer from anxiety, depression and addiction problems. Kasser (2011) concludes that, where growth of consumption is a key goal of a nation, universal psychological needs are undermined.

Table 2.1   Core universal human needs in three theories

	Theory	Core elements		
Doyal and Gough (1991)	Basic needs	Participation	Health	Autonomy
Nussbaum (2000)	Central human functional capabilities	Affiliation	Bodily integrity	Practical reason
Ryan and Deci (2001)	Psychological needs	Relatedness/ belonging		Competence Autonomy

Indeed this syndrome becomes self-reinforcing as many people turn to money and possessions as a way of coping with distress.

Another contender is the capability approach of Martha Nussbaum, who recognises the existence of universal 'human functional capabilities'. Initially, she derived this concept from neo-Aristotelian reasoning, but subsequently following extended field research in India based it on the Rawlsian idea of an emerging 'overlapping consensus'. It is notable that more recently Nussbaum relies ultimately on the language of 'need' (Nussbaum 1993, 2000, 2006; Brock 2009; Gough 2014). She identifies ten universal functional capabilities, but regards three as core: affiliation, bodily integrity and practical reason.

Table 2.1 demonstrates the close agreement on core human universals between these three theoretical approaches.

Human Needs and Sustainable Wellbeing

Universal human needs have (at least) six theoretical features that aid us in identifying *sustainable* wellbeing. They thus offer help in interpreting intergenerational claims such as arise from climate change.

First, human needs are *objective*. The truth of the claim that a person needs clean water or some minimal level of security in childhood depends on the objective physiological and psychological requirements of human beings and the nature of the satisfier, including its capacity to contribute to the health and autonomy of the person. In contrast, the truth of the claim that a person prefers Bowie to the Beatles depends on the nature of the person's beliefs about and attitudes towards the objects. Put another way, statements about wants are intentional, whereas statements of need are 'extensional': their truth depends on 'the way the world is' and not 'the workings of my mind' (Wiggins 1987).

Second, human needs are *plural*; they cannot be added up and summarised in a single unit of account. In addition to the basic needs for health and autonomy, we identify a list of universal *intermediate needs*: water and nutrition, shelter and energy, a non-threatening environment and work practices, significant primary relationships, security in childhood, physical and economic security, education and healthcare. Other need theories arrive at similar lists. Sabina Alkire (2002, ch. 2) surveys over 30 lists of 'dimensions of human fulfilment' and demonstrates a broad overlap of components. It is notable that all such lists include not only material goods but psychological goods, activities and relationships.[3]

Third, needs are *non-substitutable*: one domain of need satisfaction or objective wellbeing cannot be traded off against another. More education is of no immediate help to someone who is ill through lack of vitamin C. Thus certain *packages* of need satisfiers are necessary for the avoidance of harm. This is quite different from consumer preferences in economic theory, where substitutability is the default assumption: given a bundle of two goods it is always possible – by reducing the amount of one fractionally and increasing the amount of the other fractionally – to define a second bundle between which a consumer is 'indifferent' (O'Neill 2011).

Fourth, needs are *satiable*. It can be shown that the amount of intermediate needs required to achieve a given level of health and autonomy diminishes as their quantity increases, eventually plateauing (Dietz et al. 2009; Jorgenson 2014). Thus the contribution of calories, dwelling space, even levels of childhood security, to basic needs can be satiated. In the case of the basic needs of health and autonomy, *thresholds* can be conceived where serious harm is avoided such that acceptable levels of social participation can take place.

Fifth, needs are *cross-generational*. This is of great importance, since global warming will progressively impose dilemmas of intergenerational equity. We can assert with much confidence that the basic needs of future generations of humans will be the same as those of present humans. To avoid serious harm and to participate and act within future human societies people will require the same preconditions: not just survival, but health and autonomy. Future people will have needs for affiliation, health, cognitive and emotional expression, understanding and critical thought. The epistemology of reasoning about needs remains extensional, not intentional, and thus avoids the indeterminacy of reasoning about future preferences. Until the genetic make-up of *Homo sapiens* changes

---

[3]   The presence of various aspects of 'security' in this and other lists is noteworthy. Satiating needs at one point in time ignores temporal life spans. The *stability* of need satisfaction is critical; it makes no sense to say 'Yes, I'm fine, thanks. Filing for bankruptcy tomorrow' (Jackson 2009).

significantly, our successors will need specific amounts of the full range of basic and intermediate needs.

Together, this amounts to a remarkable – and pretty obvious – degree of knowledge about the constituents of future people's wellbeing. Compared to the indeterminacy of future generations' preferences or happiness (or of Sen's capabilities), a theory of need provides some firm foundations on which to build sustainability targets for public policy. The abilities to provide these components of objective welfare should be safeguarded for future generations. In O'Neill's (2011: 33) words: 'Each generation needs to pass down the conditions for livelihood and good health, for social affiliation, for the development of capacities for practical reasoning, for engaging with the wider natural world and so on.'

Finally, human needs have a sound ethical grounding that preferences do not: they come along with claims of justice and equity in tow.[4] Claims of need make moral demands on agents that preferences do not (O'Neill 2011). Universal needs imply ethical obligations on individuals and claims of justice – universal rights and obligations – on social institutions. An important corollary of the moral import of human need is that meeting needs should be given priority over meeting wants whenever the two conflict or if resources are scarce. Human needs, present and future, trump present (and future) consumer preferences. Need satisfiers have priority over surplus goods. More is said about this below.

Need Thresholds and Shortfalls

But what is an 'acceptable level' of need satisfaction? If need theory is to guide public policy, whether environmental, social or economic, some methodology is required to define need *thresholds*. All measures of sufficiency entail specifying a target level and then measuring the shortfall between observed levels and this target. Thus health shortfall indices measure the health gap, poverty levels measure the poverty gap, and so on (Ruger 2009). The question then arises: how is the target level set? Our answer can be divided into three stages.

First, the only morally relevant threshold for *basic* need satisfaction is the *optimum* level. 'In principle, [need] satisfaction is adequate when, using a minimum amount of appropriate resources, it optimises the potential of each individual to sustain their participation in those constitutive activities

---

[4] 'For standard economic analysis everything is a preference: the epicure's wish for a little more seasoning, the starving child's wish for a little water, the collector's wish for one more painting, and the homeless person's wish for privacy and warmth, all are preferences' (Shue 1993: 55).

important for furthering their critical interests' (Doyal and Doyal 2013: 14). Nussbaum (2011) endorses something very similar: 'human dignity requires that all citizens exceed an ample threshold level of ten central capabilities as a bare minimum'.

Second, at the level of *intermediate* needs, a different threshold applies. Intermediate needs are those properties of goods, services, activities and relationships that enhance health and autonomy in all cultures. This enhancement generally follows an asymptotic curve: additional increments of an intermediate need generate decreasing increments of basic need satisfaction until at a point no additional benefit is derived. We can call this point the *minimum optimorum* or *minopt* threshold: the minimum quantity of any given intermediate need satisfaction required to produce the optimum level of basic need satisfaction. In principle, this defines threshold levels for each intermediate need.[5] In practice there are certain difficulties, notably in the area of healthcare where huge resources can secure only marginal improvements in health, and to a lesser extent in education and economic security. But in principle the method for identifying a threshold for sufficiency is clear (Doyal 1995; Ruger 2009).

However, third, the real world is characterised by vast differences in socio-economic resources between nations and peoples, complicating any agreement on acceptable levels of wellbeing. In a transitional phase towards a more just distribution of global wellbeing, it is common to suggest a *constrained* optimum threshold for low- and middle-income countries. This identifies what *in practice* the best-performing countries at any given level of development have achieved. For example, Costa Rica regularly heads social and environmental indices among middle-income countries. Clearly this option does not and should not justify complacency and rule out necessary social change, but it does establish meaningful and justifiable transitional welfare thresholds, as we show in Chapter 4. Cross-generational thresholds for *sustainable* welfare pose additional questions, discussed below.

NEED SATISFIERS AND THE DUAL STRATEGY

While basic needs are universal, they are satisfied in countless different ways, which vary across cultures and times. Max-Neef (1989) usefully distinguishes needs from *need satisfiers*: the goods, services, activities and relationships required to satisfy needs in any given social context. For

---

[5]   This 'vitamin model' was first used to consider the diminishing impact of additional vitamins on human health. It has been extended to other health inputs and beyond that to competence and autonomy (Warr 1987; Doyal and Gough 1991).

example, the needs for food and shelter apply to all peoples, but there are wide varieties of cuisines and forms of dwelling that can meet any given specification of nutrition and protection from the elements. It is essential to draw a sharp distinction between universal needs and specific satisfiers. Without it, need theory could justly be accused of being paternalist and insensitive to context, culture and time.

How can a social group identify needs and appropriate need satisfiers? In the face of radical disagreements over the perceived interests and needs of different people, can we identify *any* bases for consensus here? To identify actual need satisfiers in any particular context or culture requires a quite distinct social process of understanding and research that brings together two forms of knowledge (this draws on Habermas (1987) and Rawls (1971), cf. Doyal and Gough 1991, chs 7, 14).

First, there is *codified* knowledge embodied in the knowledge of practical experts, whether in health and medicine, science, engineering, biology, technology or policy science. This knowledge is commonly used to help identify 'what people need': the components of a healthy diet, the educational needs of children, the factors that contribute to air pollution, and so on. Experts have a vital role to play in identifying need satisfiers (notwithstanding the disdain increasingly heaped on experts in our post-modern landscape). Needs differ from preferences partly because of this appeal to an externally verifiable stock of accumulated, yet continually contested, knowledge.

Second, there is *experientially grounded* or *practical* knowledge – the entire range of understandings, wisdom and accumulated problem-solving of people in their everyday lives and contexts. This too must contribute to deciding what objects, activities and relationships are necessary or essential for wellbeing in any given context. Public programmes that ignore this input can be irrelevant, inefficient, stupid or oppressive; one well-documented example is the construction of new housing estates far from employment, shops or social activities without public transport and concentrating poor families within them.

Clearly there is a dilemma here. Lauding the indispensable knowledge and power of science, technology, professions and experts runs the obvious risks of uniformity, inflexibility, paternalism, disempowerment and domination, threatening cultural integrity of groups and the autonomy of individuals. On the other hand, initiatives to utilise experiential knowledge, to engage and empower the powerless, can carry the danger that sectional and short-term interests coupled with power differences in a context of inequality and media concentration can undermine the identification of longer-term generalisable interests.

Thus we conclude that any rational and effective attempt to resolve disputes over need satisfiers must bring to bear *both* the codified knowledge of experts and the experiential knowledge of those whose basic needs and daily life world are under consideration. It requires a *dual strategy of social policy formation* which values compromise, provided that it does not extend to the general character of basic human needs and rights. (Doyal and Gough 1991: 141)

Inevitable disagreements over appropriate need satisfiers must be confronted and resolved in a forum as open, as democratic, and as free of vested interests as possible. What is required is a form of *procedural rationality* (Funtowicz and Ravetz 1994; Gough 1994, 2000). Determining need satisfiers entails 'a problem-solving process rather than a preference-aggregating one' (Özkaynak et al. 2012).

This is easier said than done, especially when debating need satisfiers for sustainable consumption in a world driven by economic growth and material consumption yet faced with the profound complexities and uncertainty of climate change. It would entail forms of extended dialogue and consensus-building in public forums at different levels of decision-making. Most present initiatives such as citizens' juries are exercises in consultation, not proper participatory decision-making. Some ways must be found of scaling up these initiatives while at the same time attending to power differences and distortions of debate if sustainable need satisfiers are to be identified even in Western representative democracies. For example, people's assemblies (or citizens' forums or juries) could be expanded, where lay members consider evidence and discuss relevant questions with experts, as well as amongst themselves. Their findings could then be presented to and negotiated with councillors and parliamentarians, aiming to arrive at a broad consensus (Coote 2015). These issues are returned to in Chapters 7–9.

MATERIAL SATISFIERS AND SUSTAINABLE WELLBEING

Meeting human needs also requires a socio-economic system that produces and distributes the necessary and appropriate need satisfiers – and ensures that all this does not threaten planetary limits. We identify four societal preconditions – production, reproduction, cultural transmission and political authority – which have to be satisfied by all social groups if they are to survive and flourish over long periods of time (Braybrooke 1987; Doyal and Gough 1991, ch. 5). These societal preconditions are shown at the bottom level of Figure 2.1. It is important to conceptualise these preconditions separately from intermediate needs. The former are attributes of individuals, the latter of collectivities. There exist critical *institutional* satisfiers alongside the intermediate needs.

Elsewhere, I discuss at length the material and procedural preconditions for improving need satisfaction and attempt to evaluate the strengths and weaknesses of alternative economic institutions (Doyal and Gough 1991, chs 7, 12; Gough 2000, ch. 2). Here I discuss solely the idea of *sustainable* preconditions. 'Sustainability' is not easy to define; perhaps the most generic meaning is 'that whatever is being considered has the capacity for continuance' (Ekins 2014: 56). I have argued above that need theory enables us to specify, with considerable confidence, what aspects of human wellbeing should have the capacity for continuance through time. Now we need to spell out in more detail what this means for a sustainable economy and society.

Sustainability is often understood in three domains: economic, social and environmental, all of which are important for the arguments in the book. Each requires provision to be made for their 'continuance'. Economics recognises the need for a stock of physical capital to be replaced to provide a future stream of goods and services. This requires a measure of the depreciation of capital, which will enable the calculation of the net capital stock and so net income. (It is astonishing that the ubiquitous measures of *gross* domestic product or *gross* national income still flout this basic rule.)

However, it is clear that future benefits flow from sources other than manufactured capital, sources located in the social and environmental domains. In the last two decades this recognition has been reflected in the notions of 'human capital' and 'social capital' on the one hand and 'environmental' or 'natural' capital on the other.[6] Since climate change will impact most directly on natural capital, I begin with that.

From the perspective of human welfare, we can identify certain 'environmental functions' that provide goods and services that satisfy human needs: resources for the human economy, sinks for its waste products, and crucially the entire range of 'ecosystem services' that support human life and welfare (Jacobs 1995). Environmental sustainability then requires, according to Ekins (2014: 57):

> the maintenance of important environmental functions and the natural capital which generates them. Important environmental functions may be considered to be those that are not substitutable, those whose loss is irreversible and is likely to lead to 'immoderate' losses ... and those that are crucial for the maintenance of health, for the avoidance of substantial threats (such as climate stability) and for economic sustainability.

---

[6]   This substantial literature is summarised in Gough et al. (2007).

The natural capital that performs such functions can be called *critical* natural capital.[7]

To identify critical natural capital the distinction between weak and strong sustainability is important. *Weak* sustainability assumes that natural capital is *substitutable* with physical or human capital – if climate change denudes natural capital and its functions, these can be compensated for by more investment in built capital or human capital. It suggests that natural capital can be valued in money terms, with the implication that public policy should focus on maintaining the total money value of the capital stock.

Strong sustainability rejects this and holds that the depletion of certain forms of natural capital cannot be compensated for by investment in other forms of capital (Neumayer 2013). This is especially so for the waste absorption functions of the environment and the range of life support 'services' supplied by the planet. 'An assumption of weak sustainability underpins the standard economic approach to decision-making that involves trade-offs between the goods and services provided by different forms of capital' (Ekins 2014: 60). This standard approach is clearly inadequate where there are non-marginal, irreversible changes, or great uncertainty over the probability of certain outcomes, or a lack of agreement about valuing human life and welfare.

An assumption of strong sustainability is important because it supports a more precautionary approach to managing climate change. It entails disaggregating different aspects of natural capital and environmental functions, and this means rejecting monetary measures in favour of direct physical indicators, such as GHG emissions. This does not mean that all natural capital is non-substitutable, a position that Daly (1996) labels 'absurdly strong sustainability'. But to assume wholesale substitutability is wrong and dangerous. The default assumption should be that of strong sustainability. Thus sustaining future wellbeing requires protecting and sustaining critical natural capital and the swathe of environmental functions it provides.

There is a strong parallel here with the discussion of substitutability of needs and preferences above. Just as consumer theory assumes substitutability between different satisfiers, such that the consumer will be indifferent between two combinations, so the weak sustainability principle of capital allows for substituting between manufactured and natural capitals. Similarly, just as need theory contends that different domains of need satisfaction are not substitutable, so strong sustainability theory doubts

---

7   To begin with it is helpful to situate the analysis in this chapter firmly at the global level. There will be numerous examples of critical natural capital pertaining to particular habitats, localities and cultures (O'Neill 2015).

that destroyed natural capital can be compensated for by more human-made capital (O'Neill 2015).

Strong sustainability then imposes a more demanding audit of social institutions than does conventional economic theory. For wellbeing to be sustained over time, a rich nexus of qualitatively different, incommensurable institutions must be passed on to future generations, a stipulation far more demanding than putting aside a certain quantum of malleable 'capital' (De-Shalit 2005).

This raises some difficult practical issues: it suggests that the pursuit of universalisable need satisfaction in the Anthropocene requires some form of cross-generational dialogue. In place of either total ignorance about future wellbeing or the imposition of current views about wellbeing on future generations, we need to recognise that there can be 'an ongoing dialogue about the nature of the good life that crosses generations' (O'Neill 2015). Of course, that is impossible with distant generations, but to think one generation ahead is conceivable and sufficient: the process can then be repeated by the next generation, and so on. Following the rapid growth of life expectancy, four generations can coexist in affluent societies today, and it is not impossible to reason about the needs of the fifth. A variety of institutions are now emerging to represent the interests of near-future generations, such as Finland's Committee for the Future.[8] In 2015 Wales passed the Well-being of Future Generations (Wales) Act and has appointed its first Future Generations Commissioner.

MAPPING SUSTAINABLE WELLBEING AND ITS SOCIAL FOUNDATIONS

We now have two concepts with which to think about sustainable wellbeing: universal human needs, and sustainable preconditions for satisfying those needs. Both concepts are measured by disaggregated, non-monetary indicators. Can we operationalise them so they can inform policy? The previous chapter considered the merit of using the SDGs to delineate the inner boundary of social wellbeing. It is useful now to compare the SDGs with the need approach and demonstrate how the theory of human need (THN) can provide a more rigorous map of the 'social dimension' and of sustainable wellbeing.

Table 2.2 places each SDG goal within the THN heading that seems most appropriate. This presentation makes clear, first, how the SDGs muddle up the distinction between needs and societal preconditions,

---

8　See http://www.euro.who.int/__data/assets/pdf_file/0005/302873/Intergenerational-equity-briefing.pdf?ua=1.

Table 2.2 Universal needs, preconditions and sustainability: clustering the SDGs

THN	THN	Sustainable Development Goal
Basic needs	Physical health	Goal 3: Ensure healthy lives and promote well-being for all.
	Mental health	Goal 3, especially 3.4: *Promote mental health and well-being.*
	Cultural understanding/ teachers	Goal 4: Ensure inclusive and equitable quality education and promote lifelong learning opportunities for all.
	Critical autonomy	?
Intermediate needs	Nutritional food	Goal 2: End hunger, achieve food security and improve nutrition.
	Clean water	Goal 6: Ensure availability and sustainable management of water and sanitation for all.
	Protective housing	*Goal 11.1: Ensure access for all to adequate, safe and affordable housing and basic services, and upgrade slums.*
	Non-hazardous work environment	*Goal 3.9: Substantially reduce number of deaths and illnesses from hazardous chemicals and air, water and soil pollution and contamination.*
		Goal 8.5: Promote decent work for all.
	Economic security	Goal 1: End poverty in all its forms everywhere.
	Appropriate healthcare	Goal 3: Achieve universal health coverage, including reproductive healthcare.
	Appropriate education	Goal 4: Ensure inclusive and equitable quality education; promote life-long learning opportunities.
	A secure childhood	*Goal 16.2: End abuse, exploitation, trafficking and all forms of violence against and torture of children (plus references elsewhere to girls only).*
	Significant relationships/ social affiliation	?
	Physical security	?

54

Societal preconditions	Procedural, material and distributional preconditions	Goal 5: Achieve gender equality and empower all women and girls.
		Goal 7: Ensure access to affordable, reliable, sustainable and modern energy for all.
		Goal 8: Promote sustained, inclusive and sustainable economic growth, full and productive employment and decent work for all.
		Goal 10: Reduce inequality within and among countries.
		Goal 16: Promote peaceful and inclusive societies for sustainable development, provide access to justice for all and build effective, accountable and inclusive institutions at all levels.
		Goal 17: Strengthen the means of implementation and revitalise the global partnership for sustainable development.
	Sustainability preconditions	Goal 9: Build resilient infrastructure, promote inclusive and sustainable industrialisation and foster innovation.
		Goal 11: Make cities and human settlements inclusive, safe, resilient and sustainable.
		Goal 12: Ensure sustainable consumption and production patterns.
		Goal 13: Take urgent action to combat climate change and its impacts (recognising the primary role of UNFCCC).
		Goal 14: Conserve and sustainably use the oceans, seas and marine resources for sustainable development.
		Goal 15: Protect, restore and promote sustainable use of terrestrial ecosystems, sustainably manage forests, combat desertification, halt and reverse land degradation and halt biodiversity loss.

Note: Italics indicate sub-goals.

55

whereas THN makes a sharp distinction between them. There is a strong case for separating out concepts and measures pertaining to individuals and concepts and measures pertaining to collectivities. Examples of the first are nutrition, health and education; examples of the second are gender equality, sustainability, inclusiveness and peace. The latter cannot apply to an individual; the former can. Of course measures of need satisfaction can be aggregated to assess the wellbeing of populations, but their fundamental unit is the individual or individuals in households.

Three additional conclusions can be drawn from this comparison. First, there is a considerable and welcome overlap between our intermediate needs and these proposed SDG goals; but this does not stretch to the less 'material' needs. *Material* needs can be defined as those that cannot possibly be satisfied without some level of material throughput in the economic system, whereas *non-material* needs can conceivably be satisfied without any extra material throughput (Jackson and Marks 1999). It is noticeable that the international community focuses almost entirely on material needs. The SDGs omit vital components of human wellbeing, like physical security, social affiliation and critical autonomy. Yet these, or closely related concepts, figure in all philosophically based accounts of needs.

Second, and on the plus side, great attention is paid to the societal preconditions for human betterment. This is important and praiseworthy. It fleshes out in some detail a current global consensus on procedural, material and sustainability preconditions for human wellbeing. But several of the SDGs do not find a parallel in need theory. For example, SD8: 'Promote inclusive and sustainable economic growth, employment and decent work for all' lumps together important need-related goals – participation in work and acceptable conditions in work – with economic growth, a questionable means to achieving these goals.

Third, Table 2.2 remains a one-dimensional portrayal of human needs and the SDGs. As I have argued, all needs require to be satisfied for a flourishing life. Proper nutrition contributes to good health, which contributes to social participation, which contributes to competencies, which contribute to economic security, and so on in a complex network. What is more, the satisfaction of these intermediate needs depends on the presence of a set of societal preconditions, as indicated in Figure 2.1. Human wellbeing requires a flourishing network of individual capabilities and institutional prerequisites.

The same can be said for the SDGs, though these are the product of political negotiation rather than theoretical deduction. To grasp the interconnections between the 17 SDGs, let alone the 167 targets, requires not a table but a network analysis (WHO 2015; Costanza et al. 2016). Chapter 4 returns to these issues.

THE MORAL SIGNIFICANCE OF HUMAN NEEDS

Ethical Obligations

Claims of need make ethical demands on people. If unmet needs mean severe harm or an exclusion from social life, then they imply a strong moral obligation to relieve that suffering and meet the basic needs that enable that participation. Moreover the universal nature of human needs widens the sphere of moral obligation to the global level and beyond that to the needs of future generations.

Within parts of the developed world, there are established 'welfare states' that recognise collective obligations to meet the basic needs of citizens for health, education, a minimum income and so on. Though they are subject to constant critiques and counter-movements, there is within them a patchy social obligation to meet the needs of strangers, whose unmet needs we do not directly witness and can do nothing individually to satisfy. However, these obligations end at national borders, and large flows of refugees and economic migrants are today questioning who counts as 'strangers' and placing new strains on national welfare systems.

Climate change globalises ethical obligations in a new and startling way. In our inherently interconnected world, such a commitment to meet the needs of strangers and to support the necessary welfare structures cannot stop at the arbitrary borders of any particular nation state. It lends powerful support to ideas of cosmopolitanism, which see the entire world as a potential moral and political community, however stony the road from here to there appears at present (Held and Hervey 2011). And it goes further, recognising that we also have obligations to protect future generations against serious harm if such harms can be reasonably predicted (Shue 2014). We have duties to ensure that the global life support system is not so damaged that it threatens the capacity of future peoples to meet their basic or intermediate needs. This is the essence of the Brundtland definition of sustainable development.

Finally, an important corollary of the moral import of human need is that meeting needs should be given priority over meeting wants if the two conflict or if resources are scarce. This can be stated formally, following Dobson (1998), who identifies four objects of concern for policy:[9]

---

[9] Dobson actually includes two further concerns: present- and future-generation 'non-human needs', which accord intrinsic status for the preservation of Nature, irrespective of its contribution to human welfare. As argued above, that is not the position adopted in this book. A major division runs within ecological thought: between those who value Nature *intrinsically* – for itself – and those who value Nature because it is *instrumental* for human welfare. The former would not give priority to humans over other organisms, indeed might

- Wp: present-generation human wants;
- Np: present-generation human needs;
- Wf: future-generation human wants;
- Nf: future-generation human needs.

The implied priority rule for the need theory set out above is:

$$Np = Nf > Wp/Wf$$

Human needs, present and future, trump present (and future) consumer preferences. This has implications for necessary goods and social policy, discussed later. In Shue's words, 'It is not equitable to ask some people to surrender necessities so that other people can retain luxuries ... The costs ought to be partitioned ... into costs that impinge upon necessities for the poor and costs that only impinge upon luxuries for the wealthy' (Shue 1993: 56).

Needs and Rights

Some writers contend that these arguments place strong obligations on individuals – notably affluent individuals – to act as moral agents with a duty to exercise self-restraint and pursue frugality. But, in a closely integrated world economic system with global threats to human wellbeing stemming from climate change and environmental degradation, there are, in Gasper's (2012) words, 'innumerable agents and innumerable victims'. Altruistic behaviour, however desirable, cannot be sufficient.

Thus global and national *institutions* will have a central role to play as main bearers of the duties stemming from climate change. A moral political economy ensures that the priority rule set out in the equation above is followed as closely as possible. This connects with issues of social justice and human rights (Rawls 1971; Doyal and Gough 1991, ch. 7; Pogge and Horton 2008; Pogge and Moellendorf 2008).

Universal needs provide an essential grounding for appeals to human *rights*: moral or legal claims possessed by 'right-bearers' that corresponding 'duty-bearers' must take seriously. It is usual here to distinguish 'negative' civil and political rights from 'positive' socio-economic rights.

---

give priority to 'non-human needs' over human needs, now or in the future (Dobson 1998; Barry 2002; Spash 2012). Such a 'deep green' approach has much to offer, but the focus of this book is precisely on the impact of climate change on the human and social dimensions of wellbeing. For this reason, I do not prioritise 'the value of Nature' over the value of human needs.

The former entail a duty of forbearance and protection, for example rights to freedom of expression and against discrimination; the latter entail a duty of assistance and provision, for example rights to education or healthcare. Both sets of rights can be traced back to the 1948 Universal Declaration of Human Rights, and have been elaborated and specified since then. Claims that human rights must include social and economic rights have proved to have a strong mobilising effect over the last century (Blackburn 2011).

In the face of climate change, such rights also impose strong *duties* or obligations on both agents and institutions. This raises big philosophical questions, but among scholars sympathetic to arguments of universalism three climate-related obligations are commonly distinguished: first, to set a planetary emissions ceiling; second, to allocate this quantum of emissions fairly between nations and peoples; and, third, in the case of unpreventable continuing climate change, to fund adaptation and compensation programmes for those groups most affected (references in Gough 2015a).

The first obligation, supported by almost all ethical principles, is to curb dangerous levels of GHG emissions that threaten severe, potentially catastrophic harm to future peoples and to their ability to meet their needs and pursue their basic interests. This obligation must primarily be fulfilled by institutions – notably, but by no means only, national governments. One example of this is the agreement by 195 nations at the Paris 2015 meeting of the UNFCCC to achieve 'zero net emissions' in the second half of this century.

Second, this quantum of emissions should be allocated fairly between present nations and peoples. Arguments for global distributive justice are derived from several normative approaches, but many entail a clear notion of human needs and their distinction from wants. Meeting people's basic needs would entail a principle of *moderate sufficiency*, according to which it would not be permissible to promote the less basic interests of members of the present generation if this would compromise the needs of future generations (Gough 2015a). In the Anthropocene 'meeting people's basic needs should be the *first* priority of justice':

> Since protection from harm is a matter of basic need, and since significant climate mitigation can be accomplished without compromising the *needs* of present persons, climate policy is an urgent priority of justice . . . Where our present activities are not necessary for satisfaction of present fundamental needs, and put at risk the basic needs of future generations, then they are unjust. (Wolf 2009: 355, 373)

This still leaves a third ethical question: in the face of unavoidable continuing climate change, and the enduring hardships it will impose, who has the obligation to fund adaptation and compensation programmes for the groups affected? Here also there is considerable agreement, and there are basically three answers to the question: those who have enjoyed the fruits of energy consumption and imposed the global burdens of emissions in the past; those who have the greater ability to pay; and those least likely to be plunged into deprivation and unmet basic needs as a result. These are distinct moral arguments, but they all converge in practice when considering global justice: in today's world of great inequality between nations, the costs should be borne by the rich countries of the North.

This raises finally a parallel claim of domestic justice *within* nations: between rich and poor social classes within nations in the North – and within the East and South. This does not conflict with the obligations above; rather it endorses the cosmopolitan starting point that the individual is the basic unit, not nation states, and that all individuals count equally. Chapter 4 discusses global approaches that take this into account.

Thus there is a clear moral basis for limiting the present and future harm likely to be caused by climate change, and for distributing fairly the burdens and benefits of such programmes between peoples and nations both globally and over time. Practically, this endorses a policy of 'contract and converge', with strict obligations on rich countries to cut their emissions fast. In addition it recognises the explosion of intra-national inequality: that the rich and affluent classes are expanding fast in the East and South as well as the North – and that poor people remain in the North as well as the East and the South. Intra-national social justice remains as important as ever. Finally, our distinction between needs and wants, between need satisfiers and want satisfiers, can play a useful role in guiding decisions about what must be restrained now in order to permit future peoples to avoid harm and to flourish.

Sufficiency and Equality

A needs approach results in advocating a principle of *sufficiency* – the ethics of *enough* (Barry 2012, ch. 5). But does sufficiency trump *equality*, as Frankfurt (1987: 21) contends? 'What is important from the point of view of morality is not that everyone should have the same but that each should have enough.' The need theory outlined so far says nothing about the distribution of resources *above* the sufficiency level. Is this appropriate in a world of egregious inequality? Do not concerns of justice extend beyond the minimum? In short, is enough enough (Brandstedt 2013, ch. 4)? I conclude this chapter with four brief arguments that a needs-based approach implies a goal of not perfect equality but much less inequality than at present.

First, the dilemmas of climate change prompt a powerful justification for more equality. If, owing to past industrialism, population growth, environmental degradation and climate change, we can achieve less than optimal generalisable satisfaction of basic needs, then so be it. We will be forever living in a world of constraint. But in such a world the more equal distribution of basic goods assumes still greater moral force. Evidence that inequality itself propels higher emissions buttresses this argument (see Chapter 3). The environmental crisis renews and reinvigorates the older case for egalitarianism.

Second and related, the current climate context is characterised by serious uncertainty and a possibility of catastrophe. In this case 'sufficiency' or 'enough' provides a more precautionary principle than optimisation or maximisation, which would risk business as usual, gross inequality and overshooting planetary boundaries (Brandstedt 2013).

Third, autonomy entails certain social bases of self-respect that rule out discrimination on the grounds of sex, race, religion and so on (Nussbaum 2000, 2006). This establishes a base of equality across many human attributes – but it excludes inequality based on wealth, income and class, even though these drive up unsustainable aspirations. Nor does it apply to our intermediate needs; sufficiency in housing does not imply equality of housing (whatever that would mean). Education is rather more difficult in that gross inequality may result in an absolute block on the opportunities of a person to participate in significant social roles.

Finally there are consequentialist arguments for greater equality above the sufficiency level. Meeting needs at a decent level will in practice require substantial redistribution of wealth and income. And, since inequality of wealth results in unequal power, the ability to secure the interests of the rich within state power will need to be curbed. Greater equality will be a precondition for the transformative needs-based agenda and institutions discussed throughout this book.

It follows that need theory provides a powerful case for sufficiency, prioritising needs over wants, and distributing resources more equally. These merits are all strengthened in the presence of dire anthropic pressures on the planet.

CONCLUSION

This chapter sets out an objective conception of wellbeing – human needs – and shows how it can provide evaluative guidelines for assessing prospects and policy in an era of global warming. There are four main conclusions that guide future arguments in this book.

First, human needs can be conceived and identified on a universal scale. They must be met in order for people to avoid harm and be able to function – to pursue their own goals, to participate in society and to be aware of and reflect critically upon the conditions in which they find themselves. Only if we understand needs in this way – in universal terms, applied across time and place – can we plan for and measure progress towards our social and environmental goals, both globally and into the future.

Second, needs trump wants. Needs are morally significant in a way that individual preferences are not. They accord significant rights to individuals and they impose significant obligations on individuals and institutions. These obligations and rights extend across space to embrace the global community and across time to embrace at least near-future generations of people. The preferences of consumers should not automatically stand in the way of meeting needs. Sufficiency for all trumps maximisation of utility for some.

Third, need theory demands a more specific and concrete approach to understanding the societal preconditions for sustainable wellbeing. 'Strong' sustainability imposes a more demanding audit of social institutions than does conventional economic theory. For wellbeing to be sustained over time, a rich nexus of qualitatively different, incommensurable institutions must be passed on to future generations.

Fourth, need theory can provide solid foundations for identifying the 'social' dimension of human wellbeing – the inner ring of Raworth's lifebelt. More conceptually robust than the SDGs, it can strengthen some of the thinking behind them. The theory of human need provides a normative foundation for assessing the social implications of climate change and climate policies in the Anthropocene.

The arguments put forward here for an objective, universal concept of human need do not rule out other conceptions of wellbeing. A convincing consensus is emerging that combinations of approaches – objective, subjective and relational – provide a more rounded picture of human wellbeing. For example, the New Economics Foundation defines wellbeing as a combination of how people feel and how they function, on both a personal and a social level, and has constructed *National Accounts of Wellbeing* (Michaelson et al. 2009, 2012). The OECD argues there is a convergence of understanding around non-monetary measures of wellbeing and suggests lists of indicators from developed and developing countries (Boarini et al. 2014). Statistical agencies across the world now compile 'dashboards' of wellbeing indicators. This book does not deny the usefulness of this work, especially in developing progressive policies within nations and localities. But it argues that in an era of accelerating climate change these ideas must be underpinned by a robust idea of welfare that can apply across the globe and into future time. I contend that need theory provides a more secure theoretical foundation for the numerous current empirical efforts to devise non-monetary indicators of wellbeing.

3. Climate capitalism: emissions, inequality, green growth

INTRODUCTION

This chapter turns to the contemporary world of 'climate capitalism': an economic system that aims to square the drive to growth with a decarbonising global economy. It asks the following questions: can such a combination be realised? And can it at the same time enhance human welfare and equity?

The chapter is organised as follows. It first analyses the Great Acceleration of the global economy since the Second World War and its relation to the Great Acceleration in global emissions. It goes on to discuss green growth – the strategy to decarbonise the economy whilst maintaining in broad part current global institutions and in particular the goal of economic growth. It is the dominant strategy in the world today, enshrined in the Paris agreement. Some of the inconsistencies of this strategy are demonstrated.

The chapter then returns to the social dimension by charting income and wealth *inequality* and its impact on emissions.[1] It portrays the enormous international disparities in incomes and emissions, especially when measured on a consumption basis. It goes on to chart the impact of rising 'class' inequality within countries and to expose the link with rising emissions. It shows that growth alone, even green growth, cannot eliminate poverty without laying waste to the planet: a distributive dimension is essential in its own right and to implement sustainable production and consumption.

CAPITALISM, GROWTH AND EMISSIONS

Production and Growth

Figure 3.1 charts the impact of the Great Acceleration – the global spread of the carbon economy – on greenhouse gas emissions since the Second

---

[1] Non-monetary inequalities in wellbeing are discussed in Chapter 4.

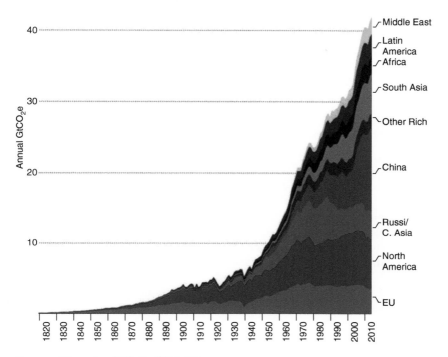

Source: Chancel and Piketty (2015: 17).

Figure 3.1 Global CO₂e emissions per region, from 1820 to today

World War. Such an epochal shift requires a political economy explanation. Its fundamental cause is the pursuit of profits and the drive to accumulate capital that shifts human economies in an epochal and irreversible way, as described in the Introduction to this book. Figure 3.1 portrays the tremendous acceleration in global greenhouse gases that has resulted.

It also shows the changing global pattern of capitalism and emissions over the past two centuries. In 1945 Western Europe and North America accounted for the bulk of production and territorial emissions. This has shrunk to 25 per cent of the total today; the Soviet Union and its successor states have followed a similar pattern. A new acceleration in the new century quickly made China the world's leading single emitter, accounting for a quarter of the current total, with South Asia and Latin America also expanding fast. Around 2004 emissions from the BRICS (Brazil, Russia, India, China and South Africa) overtook emissions from the OECD.

To explain such rapid shifts over time and patterns across countries we need to identify the second-level *drivers* of emissions. Drivers refer to 'human actions that cause climate change and the factors that shape those actions' (Rosa and Dietz 2012). They can refer to explanations of emissions over time or of cross-national variations in emissions at a point in time. The 'Kaya' identity, named after the Japanese energy economist Yoichi Kaya, provides a common way of relating four critical drivers (an identity is an equation that is true by definition):

$$CO_2 \text{ emissions} \equiv \text{population} \times (\text{GDP/population}) \times (\text{energy/GDP}) \times (\text{emissions/energy})$$

This identifies four broad factors affecting emissions over time:

- population growth;
- affluence: growth in output or income per head;
- changes in the energy intensity of output;
- changes in the carbon intensity of energy use.

Table 3.1 charts the total annual growth of global CO_2 emissions from 1971 to 2007, broken down into the four Kaya components. It shows how much the rise in emissions can be attributed to each of the four drivers, whilst holding the others constant.

Over these four decades world emissions rose fast – by just over 2 per cent a year – owing to two major factors. Population growth increased emissions by 1.6 per cent a year and growth in average incomes by nearly 2 per cent a year. But their combined effect (3.6 per cent per annum) was partly offset by the other two factors: the average amount of energy required to produce each unit of output fell by 1.36 per cent a year, and

Table 3.1   Contributions of 'Kaya' drivers to global emissions, 1971–2007

	World	OECD	China	Other NICs
Population	1.59	0.71	1.29	1.93
GDP per capita	1.96	2.07	7.51	2.84
Energy intensity	−1.36	−1.55	−4.13	−0.66
Carbon intensity	−0.16	−0.47	1.2	0.59
Net annual CO_2 growth	2.02	0.76	5.88	4.71

Notes:
In % per annum.
OECD excludes Mexico and Korea.

Source: Steckel et al. (2011, table 1).

the amount of CO_2 emitted for each unit of energy fell by 0.16 per cent a year. We can speak here of a global improvement in energy and carbon intensity. This partly reflects improvements in the emissions efficiency of production, but partly also a shift in the global economic structure, from agriculture and manufacturing to services: from now on I will refer to them together as the *'eco-efficiency'* of production.

So the world story since 1971 has been one of economic growth and population growth each driving up emissions very fast, only partially offset by improved eco-efficiency. It is this net growth of carbon that has to be halted and then reversed at dizzying speed to try to avoid 2 °C of global warming.

Table 3.1 also breaks down the results for three major regions, one corresponding to the North (the OECD) and the other two representing the East – China, and an aggregate of six other 'newly industrialised countries' (NICs): Brazil, India, Indonesia, Mexico, South Africa and South Korea. In the OECD both population growth and output growth are slow, yet they still outweigh a 2 per cent annual improvement in eco-efficiency. China has combined a dizzying rate of economic growth with medium growth of population, offset by a fast improvement in energy efficiency; yet so far this has not been sufficient to prevent an explosive rise in CO_2 of almost 6 per cent a year. The other NICs, whose total emissions are roughly equivalent to China's, reveal a different pattern: fast population and economic growth with virtually no improvement in eco-efficiency (up to 2007). The end result is that production-based emissions have soared in the East but for a different combination of reasons in China and the other six countries.

These findings undermine the 'environmental Kuznets curve' (EKC), named after Simon Kuznets. This holds that early stages of economic growth have a negative impact on environmental quality but that beyond a certain income level the environmental costs of growth level off and start to fall. There is considerable support for the theory across a variety of environmental impacts, but not for CO_2 or GHG emissions. There is clear evidence for a relative decoupling of emissions from output, indicating that rising eco-efficiency is having an impact. But the historical record to date is unequivocal: these advances are outweighed by growing production and population. There is no evidence yet of any *absolute* decoupling of carbon from output on a world scale. Indeed, there is scant evidence of any reversal of emissions growth, except in times of economic crisis, for example in the former Soviet Union countries in the early 1990s and in the West as a temporary reaction to the 1973 and 1979 oil crises and again following the 2008 financial crash.

The association is not perfect and there are many anomalies: for example, in 2008 Switzerland and the US had about the same income per head but a fourfold difference in territorial emissions per head. Clearly other factors are important. To understand drivers in richer detail, a more subtle decomposition analysis is required (York et al. 2003).[2] Such studies show that certain geographical factors are significant, notably climate: colder winters require more energy for heating. Another is the density of population: extensive, sparsely populated countries tend to have higher transport emissions. Certain demographic factors are also important, such as the age distribution of the population and the degree of urbanisation. The presence of fossil fuel reserves within a country tends to increase emissions, while hydro-electric potential reduces them. However, these studies ignore the significance of further major institutional differences between nations, economic, social and political, a subject I return to when looking at the OECD world in Part II of this book.

GREEN GROWTH

Given the dynamic of growth under capitalism, can this be harnessed to achieve a simultaneous and fast reduction in emissions? Is 'green growth' or 'climate capitalism' – a model which squares capitalism's need for continual economic growth with substantial shifts away from carbon-based industrial development – a panacea or a chimera (Newell and Paterson 2010)?

In one sense, green growth is the only game in town. The only logical alternatives are, on the one hand, that more growth per se is the solution to dealing with climate change and severe environmental threats – the Bjorn Lomborg (2010) and Matt Ridley (2010) perspective – and, on the other hand, that growth is the problem and we must move toward degrowth or post-growth. The first is advocated now by a dwindling band of believers. The second is threatening to the dominant world order and at present offers few transition routes from here to there (it is discussed in Chapter 8). But the middle ground of green growth covers a broad terrain encompassing quite incompatible positions, illustrated by quotes from two senior economists. According to Paul Krugman (2014), 'Saving the planet would be cheap; it might even be free', whereas Nicolas Stern (2015) claims 'A new energy-industrial revolution is required.' The following paragraphs try to identify the core ideas, beginning with orthodox economics.

The first premise of the green growth model is that growth is desirable. Immediately there are possibilities for vagueness and confusion. Does this mean the 'orthodox, undifferentiated' growth as measured by GDP (Barry

---

[2]   Using techniques such as STIRPAT. Research on decomposing emissions growth is expanding and can only be inadequately summarised here (e.g. Rosa and Dietz 2012; Duro and Teixidó-Figueras 2013; Lamb et al. 2014; Teixidó-Figueras et al. 2016).

2012)? Or growth in some adjusted measure that takes into account environmental losses and declining natural capital or more qualitative concepts approaching 'sustainable development'? Alternative measures exist and provide very different measures of economic success. One recent evaluation by two economists recognises the overwhelming case for adjusting GDP to take account of, at least, diminishing capital stocks and common pool resources; yet the authors are admirably honest in accepting that 'green growth' commonly entails the narrow, unadjusted definition of economic growth as expanding GDP (Bowen and Hepburn 2014). The reasons are twofold: that growth is necessary in poorer economies and that GDP is 'the core interest of policy-makers'. The first reason avoids addressing the issue of growth in rich countries in the context of egregious global inequalities, discussed below. The second is undoubtedly true at present.

The second premise of the green growth model is that growth itself, as well as human wellbeing, will be threatened if its ecological basis seriously degrades. The third premise is that untrammelled market forces create 'externalities' and 'market failures'. For example, Stern (2015) now identifies five market failures that threaten the ecological base, in addition to excessive greenhouse gas emissions – the 'greatest market failure of all'. These are: i) inadequate research development, demonstration and deployment of new technologies; ii) imperfections in risk/capital markets; iii) inadequate public networks; iv) inadequate reliable information; and v) inadequate appreciation of co-benefits.

The fourth premise is that these market failures require various forms of collective correction and steering in order to marry the pursuit of growth with climate stability. What this concretely entails is disputed. The minimal position among most neoclassical economists is that the price of carbon must be raised so that carbon-intensive activities are discouraged and low- and zero-carbon activities are encouraged. A higher carbon price, they argue, will feed its way through the economy, altering the relative prices of the plethora of goods and services to which different firms and consumers will respond and allowing them greater flexibility in deciding how to adjust. The result is claimed to be more efficient carbon mitigation. What is surprising is the insouciance with which this policy of 'getting prices right' is advocated, given the vast disruption that a realistic rise in the price of carbon would cause to current economic and social patterns of life, the interests it would challenge, and the political opposition it would generate.[3]

Some economists would now go further and call for certain forms of state

---

[3] The central estimate for an effective carbon price by the Committee on Climate Change (2016) in the UK is around $100 a tonne by 2030, rising to over $250 by 2050. Recent oil price increases have exceeded the equivalent of $100 per tonne, and in Europe taxes on gasoline are

'steering', to provide strategic direction to economic actors, both producers and consumers. If this means addressing Stern's list of market failures it amounts to a move away from the neoliberal state to 'large-scale, persistent, and pervasive collective action at various levels (whether state-led or driven by groups within civil society)' (Bowen and Hepburn 2014: 412).

'Green growth' is a strategy premised on long-term economic benefits flowing from environmental protection in general and carbon mitigation in particular. It has been advocated from at least three perspectives (Jacobs 2012). A contemporary Keynesian argument is that a green stimulus, a 'green new deal', is particularly effective in raising demand and growth in a prolonged post-crisis recession like the present. A second case revises standard growth theory to argue for the gains to growth stemming from investment in 'natural capital' and the correction of market failures. A third, based on long-wave theories of technological revolution, argues that the next industrial revolution will be based on decarbonising the economy. All versions are confident of the net benefits accruing to business and finance, including 'first-mover advantage' of firms in the technological lead, and greater security in the face of rising energy costs and climate change costs (Newell and Paterson 2010).

Green growth forms the centrepiece of the Paris Agreement, based on a resurgent *national* approach to climate mitigation following the failure of the multilateral approach at the Copenhagen COP in 2008. At COP 20 in Peru in December 2014 all countries were invited to submit 'intended nationally determined contributions' (INDCs) indicating country pledges to cut emissions after 2020. These were widely adopted, and by the time of the Paris COP almost every nation in the world had submitted an INDC. This 'bottom-up' approach broke the deadlock and led to the treaty agreement at the Paris COP in December 2015.

But the existing INDCs demonstrate how far there is to go. Global intended emissions are lower than those which would have resulted from 'business as usual' but are still much higher than the minimum of what the science agrees is necessary to stay within 2 °C by 2050. Emissions in 2030 will be less than two-fifths of the way between hypothetical business as usual and a pathway that is consistent with the 2 °C warming limit. A more pessimistic scenario predicts emissions only a *quarter* of the way (Bassi and Averchenkova 2016). These are wide shortfalls.

More detailed sectoral studies concur. Even if electricity generation could be completely decarbonised, using solar, wind and nuclear, it would not suffice because it accounts for only a quarter of global GHGs. All other

---

greater still. It is the idea of a price constraint that is both global and applicable to all carbon that amounts to wishful thinking.

sectors must reduce emissions too, which is a tall order. For example, the industrial production of bulk materials – steel, cement, plastic, paper and aluminium – accounts for about a quarter of emissions and is already very 'efficient'. Given growth, its demand is likely to double in the next 40 years, and even if its emissions efficiency could improve by a further 50 per cent this would leave the present level of emissions unchanged. There are a variety of options – recycling, substituting other materials, and carbon capture and storage – but their ability to make a significant contribution to an absolute reduction in material emissions is limited. For example, 200 kilograms of steel and 40 kilograms of cement are produced each year for every person on the planet, yet the only conceivable alternative construction materials are stone and wood, which introduce further problems. At some stage a further reduction in demand for materials must be considered. But a reduction in demand poses quite different issues to green growth and implies a major shift in the current global political economy (Allwood et al. 2013).

Various measures of 'adjusted GDP' have been proposed as alternatives to conventional GDP. These include the Index of Sustainable Economic Welfare (ISEW), the Genuine Progress Indicator (GPI), the Genuine Savings Indicator (GSI) and the System of Environmental Economic Accounting (SEEA).[4] All provide a damning critique of GDP growth as a route to sustainable wellbeing. While global GDP increased more than threefold between 1950 and 2003, global GPI per capita among the major economics peaked in 1978 (Figure 3.2). (Interestingly, this is around the same date that the global ecological footprint exceeded the Earth's bio-capacity *and* that 'life satisfaction' measures in most countries peaked.) Comparative research shows that GPI does not increase beyond a GDP of around $7000 per capita. Yet, if income were to be distributed more equitably around the planet, the current world GDP could support 9.6 billion people at this level of income. GDP fails as a measure both of wellbeing and of sustainability (Kubiszewski et al. 2013).

To summarise, 'green growth' is a strategy premised on long-term economic benefits flowing from environmental protection in general and carbon mitigation in particular. But its nature and prospects are disputed. For some, the assumed levels of future technological development and innovation are a form of 'mythic' and wishful thinking: its main function is to act as a 'floating signifier', a term amenable to forming a broad

---

[4] The ISEW starts from consumer expenditure and then adds or subtracts various monetary estimates of social and environmental impacts. The GSI starts from wealth accounts, and calculates changes in 'extended wealth' including natural resources, manufactured capital and human capital. All such approaches are premised on weak sustainability (see Chapter 2), assuming that different capitals are commensurable and can be expressed using the same numeraire.

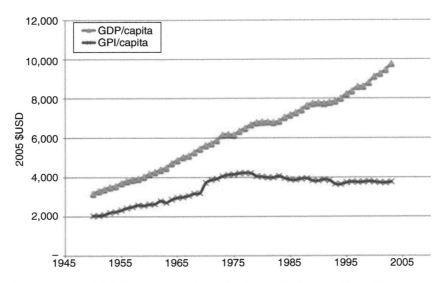

Notes: Adjusted global GPI/capita and GDP/capita. GPI/capita was estimated by aggregating data for the 17 countries for which GPI or ISEW had been estimated, and adjusting for discrepancies caused by incomplete coverage by comparison with global GDP/capita data for all countries. All estimates are in 2005 USD.

Source: Kubiszewski et al. (2013, fig. 3).

Figure 3.2   Comparing GDP and the Genuine Progress Indicator (GPI)

consensus but hiding very different interpretations (Barry 2012; Cook et al. 2012; Jessop 2012). The remainder of this chapter discusses one crucial discursive outcome of green growth – the way that it masks the role of global inequalities in income, consumption and emissions.

INEQUALITY AND CAPITALISM

Consumption and Carbon

To frame a discussion about inequality and emissions it is essential to distinguish between territorial and consumption-based or 'footprint' emissions.[5] As noted in Chapter 1, there is a widening rift between the two

---

[5] It is possible to demonstrate that given an efficiently enforced global carbon price there would be no difference between the two (Steckel et al. 2010). But a uniform global carbon price is a chimera.

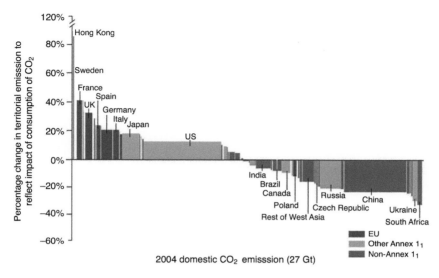

Source: Carbon Trust, in House of Commons Energy and Climate Change Committee (2012).

Figure 3.3 The impact of a consumption-based view on emissions by country

measures. Economic globalisation coupled with deindustrialisation in the North has resulted in the outsourcing of a growing share of the production to the East and notably China, which has in effect outsourced a growing share of the North's emissions. Figure 3.3 reveals this in a startling way by ranking countries according to their excess of consumption emissions over territorial emissions. Europe, the US and Japan consume goods embodying substantially more emissions than those they produce, whereas the rest of the world, and notably China and Russia, incur substantial territorial emissions which they do not consume.[6]

Figure 3.4 provides later data for the UK covering all GHGs not just CO_2. Until the financial crisis intervened, UK territorial emissions were falling but UK footprint emissions were rising; now the gap appears to widen again. The UK's place as a 'climate change leader', discussed in

[6] National footprint emissions (Ec) are calculated as follows: $Ec = Et + Ei - Ee$, where Et = territorial emissions, Ei = emissions embodied in imports, and Ee = emissions embodied in exports. The balance of trade – the last two items – will require input–output models of commodity flows across the world together with calculation of the emissions intensities of these flows. These are not easy to calculate, but great progress has been made (Hertwich and Peters 2009; Nakano et al. 2009).

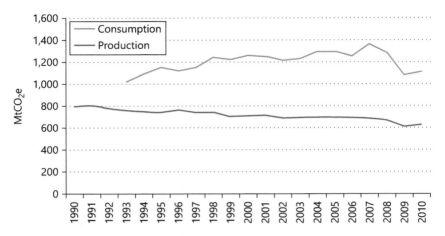

Source: Committee on Climate Change (2013, fig. 1.3).

Figure 3.4 *Greenhouse gas emissions associated with UK production and consumption 1990–2010: territorial and footprint emissions compared*

Chapter 5, rests in part on its wide trade deficit and its ability to outsource production. A report of the UK House of Commons Energy and Climate Change Committee (2012: 10) concluded: 'We are concerned that the UK could be meeting its domestic carbon budgets at the expense of the global carbon budget.'

Support has therefore been growing for the alternative 'consumption accounting principle', which measures the total emissions embodied in the consumption of inhabitants of a national territory (Bows and Barrett 2010). There are several reasons why a nation's emissions should be calculated on a consumption basis rather than a territorial basis. These will crop up throughout the book, but they can be summarised here (IPCC 2014c, ch. 4):

- Ethical: It is the people and places benefiting from consuming the goods and services emitting GHGs that should count if our interest is human wellbeing, not the people and places where these are produced. It is true that people and places where goods are produced benefit from employment, wage income and profit, but they also suffer pollution and health costs.
- Distributional: Consumption emissions are more closely related to income and wellbeing outcomes than production emissions (Steinberger et al. 2012). Measuring territorial emissions under-

estimates global inequalities in emissions – both between and within countries.

- Geopolitical: The excess burden imposed by Kyoto-style territorial accounting on emerging export economies disadvantages the latter in global negotiations over GHG targets.
- Policy relevance: Tracing consumption emissions increases the range of mitigation options by identifying the distribution of GHG emissions among different activities, imported goods, household types, and so on. This enables a better targeting of policies and voluntary actions (Bows and Barrett 2010).

The drivers of consumption-based emissions have been separately studied; they show that income level and changes in incomes are even more important than for territorial emissions. In the hypothetical absence of trade, international inequalities in consumption emissions would be lower than in our trade-driven world. Put another way, the significance of geographical, climatic and demographic factors diminishes and the role of economic factors increases when we move to consumption accounting (Lamb et al. 2014; Teixidó-Figueras et al. 2016). Doing so 'reallocates emissions from a large number of relatively poor individuals (Chinese, South Asians) to a fewer number of relatively rich individuals (North Americans and West Europeans). Taking account of consumption-based emissions thus tends to increase the level of global individual CO_2e emission inequalities' (Chancel and Piketty 2015: 28).

International Inequality and Emissions

This leads on to the role of inequality in both climate change and climate change policy. We have already encountered unequal impacts of climate change in Chapter 1. This chapter concentrates on the reverse flow: the way that inequalities in income and wealth impact on emissions and climate change. Inequality here refers to income and wealth measured in money terms; inequality in wellbeing measured by non-monetary indicators is addressed in Chapter 4. The first task is to plot inequalities in money terms between and within countries.

Inequality *between* countries widened throughout most of the twentieth century as the industrialising West pulled away from the rest of the world. It had colonised much of that world or exercised economic power to prevent the emergence of serious competition. The result was a widening gap until the late twentieth century and the turn of the millennium, when industrialisation and economic growth in the East really took off (Milanovic 2013).

Inequality *within* countries differs according to the measure of inequality used. If we take the income share of the top 10 per cent, then this fell from the start of the twentieth century until the 1970s across all Western nations for which we have data. Thereafter the share of the top 10 per cent has rebounded in the 'Anglosphere' (the English-speaking nations of the OECD) but less so in the major European countries and Japan. If we concentrate on the richest 1 per cent, then the pattern is roughly the same except that the rebound in the Anglosphere is greater still. The richest 1 per cent in the US now receives as large a share of income (18 per cent) as they did almost a century ago in the 1920s, and the UK has followed close behind. There is some data for emerging economies in the 'East' which show a similar pattern to the Anglosphere: the share of the richest 1 per cent roughly halving between 1930 and 1980, since when it has risen noticeably (Piketty 2014, figs 9.2–9.9). This reflects rising *class* inequality within countries.

Putting these two factors together we can calculate the pattern of *global* inequality: the global distribution of all incomes, as though every individual on Earth is ranked by income and then compared. In 1900 class inequality within countries was the main factor affecting global inequality. Throughout the twentieth century, as the North pulled away from the rest of the world in economic terms, the country in which a person was born grew in importance, such that by 2000 over two-thirds of global inequality was due to locational rather than class factors. In the present century this pattern is again reversing as class inequality resurges (Milanovic 2013).

Such global inequality is of course greater than that of even the most unequal nations (Milanovic 2013). The first column of Table 3.2 reveals extreme income inequality across the globe in 2013: the top 1 per cent of individuals receive almost one-fifth of global income and the top 10 per cent more than one-half, leaving 90 per cent to share the remainder. Of these the bottom half receive just over one-tenth and the poorest tenth just 1 per cent.

Global Distribution of Footprint Emissions

Two reports published in 2015 go on to calculate the overall global distribution of *consumption*-based or footprint emissions (Chancel and Piketty 2015; Gore 2015). To do this requires making some heroic assumptions, discussed in more detail in the Appendix to this chapter. One common assumption is that emissions vary step by step with incomes, over time and between people at a point in time. Alternatively, different emission elasticities can be applied. An 'emissions elasticity' of 0.9 assumes that Person A who is 10 per cent richer than Person B consumes goods and services that

Table 3.2   Global concentration shares of income and consumption-based emissions, 2013

	Income	CO$_2$ emissions	
		El = 0.9	El = 0.7
Top 1%	17.8	13.8	9.9
Top 10%	52.7	45.2	40.0
Middle 40%	36.3	41.8	44.8
Bottom 50%	11.0	13.0	15.3
Bottom 10%	1.0	1.2	1.5

Note: El = emission elasticity.

Source: Chancel and Piketty (2015, tables 7, A1).

emit 9 per cent more. An elasticity of 0.7 assumes that the richer person emits 7 per cent more. Table 3.2 above illustrates the importance of these assumptions. Figure 3.5 illustrates the findings of the Gore (2015) report, which assumes a 1.0 elasticity. It therefore records a higher share for the top decile (49 per cent) than Table 3.2 (45.2 per cent or 40.0 per cent).

But whatever the exact elasticity, the degree of inequality in emissions across the globe is extraordinary. The emissions of the top 1 per cent of high emitters is roughly the same as those of the entire bottom half of the world's population. These 60 million high emitters are overwhelmingly the 'old rich' of the West: for example, one-half are from the US, amounting to the top 12 per cent of its citizens. Such inequalities are egregious: a very rich American consuming and emitting over 150 tonnes of CO$_2$e a year can contribute 1000 times as much carbon as a very poor Zambian (0.16 tonnes) (Chancel and Piketty 2015).

Since the turn of the millennium global inequality in emissions has begun to decline: emissions of the middle 60 per cent of the world's population have grown very fast, the top and bottom quintiles less so. China and the East have grown much faster than the North, increasing the share of moderate emitters on a global scale. Since the Kyoto report of 1998, emissions have reflected the twin changes in the world economy: international inequalities have diminished but class inequalities in consumption emissions within countries have risen sharply (Chancel and Piketty 2015).

On balance this will reduce the global spread of emissions a little, but not enough to remove an irreconcilable contradiction between sustainability and ending global poverty – the first of the Sustainable Development Goals. If future patterns of global growth replicate those in the booming

Percentage of CO$_2$ emissions by world population

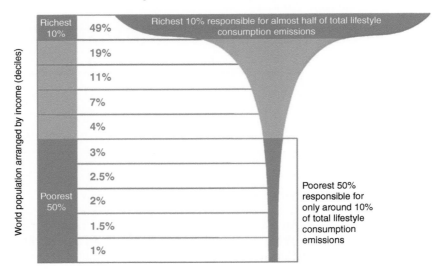

Source: Gore (2015, fig. 1).

Figure 3.5   Global distribution of consumption-based emissions

pre-crisis period 1993–2008, then extreme poverty (defined as having an income equivalent to less than $1.25 per person a day) would not be eradicated until 2077 – *60* years hence and 'beyond the life expectancy of those born today in the poorest countries'. If poverty is defined as less than $5 per person per day, which is considered appropriate to meet basic needs to a minimally acceptable level, then poverty would not be eradicated for two centuries (Woodward and Simms 2006; Woodward 2015).

There are two basic reasons for these pessimistic predictions. First, global growth since the mid-1990s has been highly inequitable: incomes of the poorest tenth of people have grown more slowly even than the average – by 1.5 per cent per annum compared with 3.7 per cent for global median incomes. Second, it builds on the extraordinary inequality already noted. Growth in the absolute increments accruing to upper income groups dwarfs those in the bottom half: the poorest third of the world population received just 1.2 per cent of the additional income generated by global economic growth between 1999 and 2008, the poorest 60 per cent received 5 per cent, while the richest 40 per cent accrued 95 per cent.

Against this background, if the business-as-usual model were used to eliminate poverty it would devastate the planet. Just to eliminate $1.25 a day poverty, global GDP would need to increase by *15 times* to an average global income of \$113 000 in 2115 (three times the top Western levels today). As Woodward and Simms (2006) argue, all strategies to eliminate global poverty are untenable unless the poor get a bigger slice of the whole cake – and the cake cannot continue to expand because of global constraints on emissions. Either new forms of redistribution or a shift to an alternative economic pathway is required.

WITHIN-COUNTRY INEQUALITY AND EMISSIONS

What will be the effect of rising class inequality on emissions and climatic stability? Does inequality push up or damp down emissions? Arguments that emissions in the Anthropocene are now driven by a growing super-rich and class inequality have given rise to the idea of the *Plutocene* (Ulvila and Wilen forthcoming).

Inequality within countries has escalated in the era of financialised capitalism since 1980 for three immediate reasons. First, the share of income from property, that is, profits, rent and dividends, has risen in the North from 15–25 per cent of national income to 25–30 per cent. This reflects the secular growth of the value of private capital over this period, a central feature of Piketty's explanation of the new 'patrimonial capitalism'. This in turn has been explained in terms of new technology, globalisation and a political counter-movement launched to free capital to pursue higher profits after the lows of the 1970s (Glyn 2006; Piketty 2014, figs 5.1, 6.5; Atkinson 2015, ch. 4).

Since income from property is more unequally distributed than income from work, this trend will automatically increase overall inequality.[7] But the distribution of wages and salaries has also become more unequal, notably in the US (continually since 1950) and the UK (since 1980). The combination of stagnant real wages at the bottom and stellar rises at the top (where earnings merge with capital income from bonuses, stock options, etc.) is well charted (Atkinson 2015). With very few exceptions (France, Japan and Spain) this pattern has been followed elsewhere in the OECD: declining shares of total income at the bottom and sharp rises at the top. Another important trend is that families with children suffer lower per capita incomes. Children are twice as likely to be poor across the world

---

[7] However, the spread of owner-occupied housing, not only in the OECD, has modified this trend in part. When the value of property in housing is imputed as a part of households' income then some middle-income groups have also enjoyed important gains.

than adults, a trend with worrying implications for future inequality and wellbeing (Newhouse et al. 2016).

Behind these immediate drivers there is general agreement that the nature of capitalism and inequality changed in the late 1970s. For a variety of reasons finance-led capitalism and neoliberalism replaced managed, Keynesian capitalism (Glyn 2006). The defining ideas included a belief in the superiority of markets and a denigration of government and collective action. Their defining characteristics included a new international division of labour, the global spread of production networks, trade and financial flows, the dominance of finance, rising profit shares – and widening class inequality (Newell and Paterson 2010; Koch 2012; Piketty 2014).

The age of neoliberalism has also been characterised by the globalisation of the fossil energy regime: a new surge in carbon emissions fuelled by global growth, population increase and new consumption norms. Now several Nobel prize-winning economists argue that inequality is bad for economic growth and stability (Stiglitz 2013). Concern about the negative effects of inequality is now central to the thinking of the International Monetary Fund, the World Bank and the World Economic Forum.

Inequality and Emissions

Does such inequality per se have an impact on emissions? Do unequal societies emit and pollute more than more equal ones? The compatibility between equity and sustainability matters for future sustainable welfare. But to discuss this question we must go beyond the macro-estimates of the previous section and delve more deeply into theory and evidence. The theoretical literature reveals no consensus on this question.

Some economists argue that inequality can actually reduce emissions, owing to the 'emission elasticity' arguments cited above (Ravallion et al. 2000). There is evidence that the marginal propensity to consume and to emit falls as income rises. This implies that higher inequality in a country would, all other things being equal, reduce consumption emissions. It could also imply that redistribution to the poor would increase emissions, since relatively more would be spent on higher-carbon necessities such as heating and food (discussed further in Chapter 7). One might speculate too that the newly emerging plutocratic class in the US and elsewhere simply have too much money to spend and consume, however many planes, yachts and mansions they own. By commandeering most of the benefits of growth in recent years they have denied poor and middle-income groups any ability to improve their consumption of necessities.

Though socially regressive and repugnant, such hyper-inequality may save on emissions![8]

A second, political economy argument claims that in richer countries inequality is associated with accelerated industrial decline, which out-sources production. This both reduces domestic emissions and weakens the power of industrial interests and unions that would oppose emission controls (Gassebner et al. 2008). This might explain the UK's leading role in legislating on climate change (Gough 2011a).

Against this, many argue that class inequality drives up emissions, for two main reasons. First, inequality increases status competition in society, an effect first noted by Thorstein Veblen in *The Theory of the Leisure Class* published in 1899, and restated with cross-national evidence by Wilkinson and Pickett in *The Spirit Level* (2009). Inequality spurs competitive consumption, emulation effects and excessive consumerism. It creates material aspirations that cannot be scaled to everyone in a sustainable manner. It fosters competition for 'positional' goods that is both counterproductive and unsustainable. More recent research shows rising inequality is associated with longer hours of work and rising debt levels, both of which stimulate consumption and emissions (see Chapter 7). Moreover, the greater the share of any growth claimed by the rich (approaching 100 per cent in some years in the US), the greater the demand for growth to provide any improvement for the 99 per cent (Bowles and Park 2005; Frank 2011; Pickett et al. 2014; Laurent 2015; Koch and Mont 2016).

Related to this, inequality hinders collective action. Higher inequality strengthens the power of the rich to make decisions, set agendas and inculcate selfish values. The rich will have the incentive and the means to substitute private amenities for public ones, which will further reduce their commitments to public provision. Others argue that inequality undermines the health, resilience and capacities of poorer communities, which multiplies the social damage following environmental shocks. Inequality downgrades environmental concerns by forcing short-term time horizons on lower-income groups. A comparison of counties in the US finds that

---

[8]   Chancel and Piketty (2015), in trying to explain the purported level of spending and emissions by the richest 1 per cent, fail to convince. Having estimated that luxury transport and household energy together might account for about $55tCO_2e$ per year, they are left with a wide gap to reach their estimate of $300tCO_2e$. Another 250 tonnes must somehow be emitted and it is difficult to see how that can be achieved.

    Against the view that hyper-inequality may save on emissions some argue that the higher savings of the prosperous and rich result in more investment that itself results in emissions, the benefit of which flows to owners of capital. But this is hard to defend for several reasons detailed in the Appendix to this chapter.

richer counties have greater consumption-based carbon emissions but lower production emissions, the latter being concentrated along with their costs in lower-income counties. More inequality thus results in more displacement: the world becomes partitioned between the 'sacred groves' of the rich and the 'sacrifice zones' of the poor – both between and within countries (Hecht 2002; Marmot 2005; Boyce 2007; Neumayer 2011; Laurent 2015; Jorgenson et al. 2016).

The theoretical literature thus arrives at no consensus, and the few available cross-national studies likewise come to conflicting conclusions or conclude that inequality has no significant effect on emissions (Jorgenson et al. 2016). However, Grunewald et al. (2012) claim to reconcile these disagreements by discriminating between countries at different income levels. Using data for 158 countries between 1980 and 2008 they conclude that, in low-income countries, rising inequality is associated with lower carbon emissions, whereas in high- and upper-middle-income countries the opposite is the case: rising inequality is associated with higher carbon emissions. The dividing line is around the income levels of upper-middle-income countries such as Mexico, Brazil, Romania and South Africa. However, this research measures production not consumption emissions.[9]

Together, these studies suggest that in the South the Ravallion effect holds: where inequality is high a large share of the population lives essentially on the fringe of the carbon economy, thus reducing overall emissions. In richer countries, research findings favour the Veblen and/or political economy explanations: rising inequality drives up consumption and weakens solidarity and collective action to restrain emissions.

CONCLUSIONS

This chapter has turned from normative arguments about need and well-being to descriptive and analytical perspectives on the global framework within which the climate crisis has unfolded. It presents a broad political economy analysis of the global capitalist economy. This is the framework within which the search for the safe and just space for humanity must perforce take place. It can be summarised as follows.

First, the vertiginous rise in energy use and emissions since the Second

---

[9] Other research using consumption-based emissions and changes over time comes to similar conclusions (Jorgenson et al. 2016). In the rich countries rising inequality lowered total footprint emissions up to 1996, but thereafter significantly increased them. In middle-income countries, positive changes in inequality were always linked to lower emissions. Among low-income countries there was no relationship.

World War, and especially over the last three decades, is fundamentally explained by the inherent drive towards the accumulation of profit and capital, which in turn drives output, incomes and consumption across the globe. At the same time, the technological dynamism of modern capitalism has driven continual improvements in productivity, both labour productivity and – later and more subdued – 'eco-productivity'. The latter can reduce the energy intensity and emissions intensity of production and thus partially – but so far quite inadequately – offset the upward surge in emissions.

Second, this belief propels the strategy of green growth, which faces so many challenges that the idea of green capitalism merits the term 'contradiction'. The chances of failing to meet climate and decarbonisation goals are high – even if one accepts a 50:50 chance of avoiding 2 °C as an acceptable goal. In addition, issues of equity and justice are sidelined, and the nature of sustainable wellbeing is unquestioned: consumer preferences and spending power continue to determine what is produced. These issues are examined more closely in Chapter 4 and Part II.

Third, the last four decades in particular have witnessed a global integration of trade, production and investment that has rebalanced the global economy away from the North towards the East. This outsourcing of production has altered the balance of emissions and widened the gap between the territorial and consumption emissions of the rich world. International inequalities in consumption footprints are wider than for Kyoto-based territorial emissions, yet the latter remain the globally agreed measuring rod.

Fourth, inequalities in income and wealth within countries have widened since around 1980 in consequence of technology, globalisation, the 'unleashing' of capital mobility and the dismantling of public regulation and provision designed to counteract this dynamic, especially in the Anglosphere. This inequality trend on balance accelerates consumer emissions in the North while adversely undermining social welfare and the collective capacity radically to reduce them. In the South inequality can prevent the poorest entering the carbon economy.

Fifth, this combination creates a tragic contradiction between growth, climatic instability and egregious inequality. All strategies to eliminate global poverty are untenable unless the shares of the poor are raised: in other words unless a more equitable model of the global economy is introduced. Redistribution and sustainable development, far from being optional add-ons, provide central policy mechanisms for counteracting global warming and protecting the environment. They are no longer simply tools of eco-social policy, but tools of planetary management. These issues are taken up in the next chapter.

APPENDIX: CALCULATING THE DISTRIBUTION OF GLOBAL EMISSIONS

The estimates discussed in Chapter 3 raise two methodological issues: what items of expenditure to include, and how to estimate their carbon or GHG emissions.

WHAT TO INCLUDE?

Though we are committed to consumption-based, not territorial, emissions, how wide do we spread the net? Consumption accounting typically attributes carbon emissions to the 'final demand' of a country and is based on the UN System of National Accounts. Thus the total 'consumption' footprint of a country does not just refer to private consumers. It refers to all components of final demand, that is:

1. private consumption by households;
2. government consumption: public expenditure on final goods and services;
3. investment (private and public) in construction, equipment, infrastructure and so on.

The term 'consumption accounting' can be used to refer to just the first or all three. In any case it is possible to calculate *government* footprints and *investment* footprints separately.

Of the two studies referred to in this chapter, Gore (2015) calculates consumption-only footprints, which account for about 64 per cent of global emissions. The argument for this is that we only have *distributive* data on household expenditures and emissions.

Chancel and Piketty (2015), on the other hand, apply the distribution of private consumption emissions across income groups to emissions from government services and investment spending. This I argue is unwarranted for both categories. First, public social services, the major item of government real expenditure (excluding transfers) in most countries, are utilised much more equitably than private spending. On average, in OECD countries, public services are worth the equivalent of 76 per cent of the post-tax income of the poorest group, and just 14 per cent of that of the richest (Verbist et al. 2012; see also Chapter 7). To allocate their emissions according to private consumer expenditure is misleading and negates the very different principles involved – need/citizenship versus money-backed preferences (Gough et al. 2011).

Regarding investment spending, there are several problems with the technique. First, to assume that savings determine investment contradicts most economic theory. Second, there is no sound methodology available to compute the ultimate beneficiaries of building a factory or buying capital equipment. Third, even if this were possible, we can't assume that richer people on average have a penchant for high-emitting capital goods and poorer people for low-carbon investment.

ESTIMATING EMISSIONS FROM INCOME OR EXPENDITURE

In the absence of direct emissions data for nearly 200 countries, the consumption footprint of different income groups is calculated by applying emissions elasticity coefficients to income or expenditure data. Both studies (Gore 2015 and Chancel and Piketty 2015) use a fixed coefficient to relate income to emissions for all countries and for all income groups within all countries. Gore (2015) assumes a 1:1 ratio of emissions to income. Chancel and Piketty (2015) provide a range of coefficients from 0.7 to 0.9, but present their main findings using 0.9; that is, if income doubles, emissions rise by 90 per cent. Table 3.2 illustrates the importance of the elasticity assumption.

However, neither study recognises robust evidence that emissions elasticity varies between countries and between income groups within countries. For example, the elasticity in OECD countries is well below 1, while that for less developed countries is not significantly different from 1 (Liddle 2015). Studies within developed countries show much lower levels, with elasticities varying from 0.6 to 0.8 for expenditure and from 0.35 to 0.52 for income. Emission levels do not keep step with rising levels of income and expenditure (Weber and Matthews 2008; Chitnis et al. 2014).

Putting these factors together suggests that the emissions are less unequally distributed across the world than are expenditures, which in turn are less unequally distributed than incomes. None of this undermines the big picture: that income (and wealth) is *the* major driver of consumption across the world and that egregious inequality results in egregious emissions inequality.

4. Sustainable wellbeing, necessary emissions and fair burdens

It is not equitable to ask some people to surrender necessities so that other people can retain luxuries . . . The costs ought to be partitioned . . . into costs that impinge upon necessities for the poor and costs that only impinge upon luxuries for the wealthy. (Shue 1993: 56)

This chapter returns to the fundamental issue of sustainable wellbeing: the idea of a safe and just space for humanity, living within planetary boundaries and providing a decent, safe and just floor of wellbeing for all people. Chapter 2 argued that the idea of universal human needs provides a robust way of identifying the inner social boundary: an aspiration of human flourishing that the whole world can and should attain. But how does this relate to the outer boundary, and in particular the urgent requirement to control and stabilise global warming within safe levels? Can a world of 9 or 10 billion people meet everyone's basic needs and still maintain safe conditions for future generations to meet their needs?

Early estimates suggested that the energy and emission costs of bringing everyone within Raworth's 'lifebelt' would be rather small. In 1985 Goldemberg et al. (1985) estimated that we could attain basic needs and much more for 1 kilowatt of primary energy per capita. Two decades later the Swiss '2000 Watt Society' advocated a per capita consumption of 2 kilowatts, corresponding to 1 tonne of CO_2 emissions per year. Other research concluded that the Millennium Development Goal targets for 2030 could be met with 1 tonne of emissions per person per year (Chakravarty et al. 2009; Steinberger and Roberts 2010). Another study of energy poverty revealed 3.4 billion people utilising less than 10 gigajoules of modern energy per capita per year (less than the energy of two barrels of oil when combusted). To raise all 3.4 billion people to the 10 gigajoules level would, at current levels of eco-efficiency, increase global energy consumption by some 7 per cent, assuming all other emissions stay constant (Chakravarty and Tavoni 2013).

These estimates are put into perspective by a World Bank study concluding that if the 40 million drivers of SUVs in the US switched to fuel-efficient passenger cars this alone would nearly offset the emissions generated in providing electricity to 1.6 billion homes that are currently without it (World Bank 2010). It is here that need theory gains its moral purchase. Driving a truck-sized auto does not constitute a necessity and, if it cannot be accommodated within the global carbon budget, it is an unjust claim on global energy and the global atmospheric commons.[1] These findings suggest that meeting human needs, unlike satisfying consumer preferences now and in the future, can provide a moral yet realist standard for sustainable wellbeing.

But other research into future energy requirements paints a different picture, suggesting that pursuing the enormous social development needs in the South and East will rapidly expand carbon emissions and could alone lay claim to two-thirds of emissions by 2050. A different conclusion follows from these studies: global emission targets are unlikely to be achieved without endangering sustainable development; or, vice versa, social development goals cannot be achieved without overstepping the available carbon space (Steckel et al. 2013). The dilemma suggested by the Brundtland Report again rears its head: how to ensure that the needs of both current and future generations are met.

This chapter explores these issues in more detail. First, I look at the inner boundary and what would constitute a moral minimum of need satisfaction across today's world. I then estimate what claims this minimum would make on available global carbon space – what 'necessary emissions' they would entail. Second, I look at conflicts and synergies between the pursuit of wellbeing and of sustainability – between the SDGs and 2 °C – and ask how the conflicts might be reconciled and the synergies nourished

The chapter, and Part I of the book, ends with two policy conclusions. The first is that green growth must be accompanied in the rich countries by a shift towards sustainable consumption. Climate policies targeting production and supply must be accompanied by policies targeting consumption and demand. The second is that sustainability and distribution are intimately connected. The burdens that accompany climate mitigation must be seen to be shared in a roughly fair way between nations and classes. This points at the international level towards a global equity framework.

TRACKING SUSTAINABLE WELLBEING

Sustainable wellbeing expresses a relationship between wellbeing and sustainability. One way to conceive this is as a *ratio* that some have called 'the ecological efficiency of wellbeing' (Dietz et al. 2009). One

---

[1]   Unless the SUV is essential for productive uses. Most SUVs are not.

well-known measure is the Happy Planet Index (HPI) developed by the New Economics Foundation in 2006 (Abdallah et al. 2009), defined as:

Happy plant index $\approx$

$$\frac{\text{(Life expectancy} \times \text{Experienced wellbeing)} \times \text{inequality of outcomes}}{\text{Ecological Footprint}}$$

The measure of wellbeing (the numerator) is a multiple of objective well-being (life expectancy) and subjective wellbeing (a measure of overall life satisfaction on a scale of 1 to 10 obtained from the Gallup World Poll) to give an indicator of *Happy Life Years*. Since 2016 this has itself been multiplied by an inequality index to assess how far the national levels of life expectancy and experienced wellbeing are dispersed across countries. The denominator, the ecological footprint, measures the average amount of land needed per head of population to sustain each country's consumption pattern.

The HPI thus brings together three factors discussed in this book – a basic need (survival/health), inequality and sustainability – in an elegant and parsimonious way. It consistently finds sustainable wellbeing to be highest in a group of countries around the Caribbean, with Costa Rica the top country again in 2016, together with some countries in South-East Asia, such as Vietnam, Bangladesh and Thailand. In most cases they rate highly on life expectancy and experienced wellbeing coupled with medium levels of inequality and ecological footprint.

The HPI has been subject to several criticisms. First, it includes a measure of subjective wellbeing, which, though enlightening and useful as a measure within countries and communities, has been much criticised as a measure of wellbeing across cultures and times (see Chapter 2). Second, the ecological footprint has been much criticised as a measure of sustainability (van den Bergh and Verbruggen 1999).[2] However, the HPI's value lies in raising awareness of the links between social, economic and environmental factors, and in stimulating debate about how to combine and measure them on a global scale.

The wellbeing approach in this book is based on a set of objective universal needs, of which life expectancy is one measure. My focus on climate change points to *GHG emissions per head* on a *consumption basis* as the main measure of sustainability. To begin with I consider the numerator,

---

[2]   There is also a technical problem: all ratio measures tend to be dominated by either the numerator or the denominator depending on which has the greater dispersion (in this case the denominator). The HPI recognises and corrects for this, but it has been found that the correction distorts the findings (Abdallah et al. 2009; Knight and Rosa 2011).

the measure of wellbeing, in more detail. There are two main approaches. The first is to measure the satisfaction of core needs – health, autonomy and participation – in terms of outcomes. Measuring autonomy and participation is difficult, so in practice attention has been focused on health outcomes, notably life expectancy. Figure 4.1 shows national consumption-based (not territorial) emissions and life expectancy for a large number of countries in 2011. Such plots always display a 'saturation curve': life expectancy rises rapidly as national incomes and emissions rise from low to medium-low but thereafter levels out.

These data point to something unexpected and optimistic: countries in the 'happy planet' position of enjoying high average life expectancy (above 70 years) while emitting less than 1 tonne of carbon – or 3.7 tonnes of CO_2[3] – per person. The inset area in Figure 4.1 shows the top left corner, what has been called 'Goldemberg's Corner', in greater detail. These countries show that meeting a basic need – survival and a certain measure of good health – can be achieved with relatively low emissions, even within present-day economies. A total of 3.7 tonnes of CO_2 per head is still far too high to restrain global warming below 2 °C, but 12 of the countries in that corner can be seen to consume less than 2 tonnes of CO_2 per head.

The bulk of these countries are medium-income with relatively closed, self-sufficient economies. A majority are in South and Central America and have warm winter climates. Others are found in South-East Asia, Europe and North Africa. The variety of countries in the corner suggest there are separate routes to low-carbon social development (Lamb et al. 2014). However, one group is notably absent: the high-income 'welfare states' of Europe and the North, none of which approach sustainable wellbeing (Koch and Fritz 2014).

Intermediate Needs

But life expectancy alone is an inadequate measure of wellbeing. Does the saturation curve, and Goldemberg's Corner, hold up for a richer range of human needs? Chapter 2 introduced the concept of universal intermediate needs and showed how these related to, and clarified, some of the Sustainable Development Goals. Researchers are now defining measures of intermediate needs at the global level and specifying minimally decent thresholds for each. 'Decent' here means above poverty-level standards but below the optimum or minopt standards advocated in Chapter 2, for example a constrained optimum threshold equivalent to the best national

---

[3]   The atomic weight of CO_2 is 3.7 times that of carbon (C).

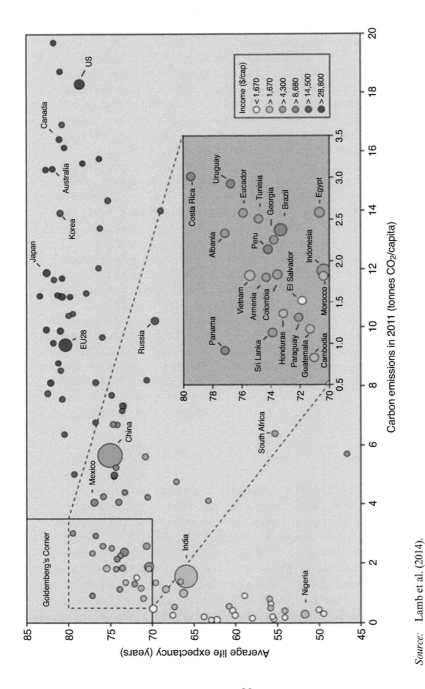

Source: Lamb et al. (2014).

Figure 4.1 Life expectancy and consumption-based emissions, with enlarged view of countries in 'Goldemberg's Corner'

standards achieved by countries at different levels of development (such as Costa Rica for middle-income countries).

To begin with, almost all studies focus on *material* need satisfiers. A wide range of human needs exist for sociability, primary relationships, some forms of security and so on, but when calculating climate-sustainable wellbeing researchers concentrate only on material need satisfiers. To define thresholds, two criteria can be used. First, there are standards advocated by the UN and other international agencies, for example for nutrition, living space, access to water, education and healthcare. The second criterion is more empirical – the spread of access to critical goods and services across all countries where they are available and affordable. For example, ownership of mobile phones is trending towards saturation (over 90 per cent) in urban areas of select emerging economies (China, India, Brazil and South Africa), and in Africa mobile phone penetration is already 67 per cent[4] (Rao and Min 2017).

But specifying threshold standards in other domains can be more problematic. For example, the recommended minimum dwelling space varies from an inhuman 1.3 square metres per person in the Millennium Development Goals to 10 square metres per person in another study based on the floor space at which middle-class Indian homes tend to plateau (Rao and Baer 2012). Similar disagreements characterise many other satisfiers, for example the amount of water per capita required for decent living conditions. Nevertheless, international standards and global consumption practices offer evidence around which some consensus may be achievable.

It should be noted that all such lists of needs at the household level entail assumptions about collective provision, notably investment and public provision. Investment and the inherited infrastructure will affect access to water, electricity and other collective goods. Public provision by governments of education and health services especially will impact on the level of individual services households are required to finance. Finally, the entire exercise makes assumptions about the shared structures of living – to what extent nuclear families or larger units occupy dwellings and how far durable goods, such as cars and washing machines, are shared.

Necessary Emissions

Using some of these need indicators, we can estimate the minimum emissions required to achieve them given current levels of eco-efficiency. One study includes measures of access to five critical intermediate needs:

---

[4]   See http://guardian.ng/technology/africas-mobile-phone-penetration-now-67/.

nourishment, water, sanitation, electricity and non-slum urban housing. Calculations of countries which supply all these measures to over 90 per cent and over 75 per cent of their populations again reveal that some countries achieve a wide spread of need satisfaction with low emissions levels. Again they are mainly middle-income countries with warm or temperate climates and low transport emissions (Rao et al. 2014).

This research paints a relatively benign picture of the potential for meeting basic material needs in a sustainable way. If Costa Rica can achieve such high levels of health and material standards with an average income of $10000 and a national footprint of less than 2 tonnes of CO_2 per person, can we not learn some lessons? It also shows that simple 'stages of growth' theories, such as the environmental Kuznets curve, cannot explain the variety of welfare–emission combinations found in the world today, despite the growing impact of globalised trade, finance, production and consumption standards. To explain the 'happy planet' countries requires drawing on a wider range of factors: history, geography, institutions, social relations, path dependency and power, as discussed in Part II of this book (Lamb et al. 2014).

But other research is more pessimistic. One study estimates the future energy and emissions required to achieve decent standards of life expectancy and infant mortality and to meet five intermediate needs (access to water, sanitation, electricity, adequate nourishment and basic education). Under several feasible assumptions about eco-efficiency this goal would require fast-rising emissions up to at least 2030 for China, centrally planned Asia and South Asia – and continuing up to 2050 for sub-Saharan Africa. Altogether this could lay claim to 63 per cent of global emissions by 2050 (Lamb and Rao 2015). Global emission targets are unlikely to be achieved without endangering sustainable development; or, vice versa, social development goals cannot be achieved without overstepping the available carbon space (Steckel et al. 2013).

More research is needed to clarify the relationship between social development and emissions and to explain these conflicting findings. These are partly due to different methodologies.[5] The openness of economies to trade and their ability to import high-carbon goods also appear to be relevant in driving up future development emissions. Continuing with current forms of globalisation and the spread of high-carbon consumption

---

[5] Researchers arguing the pessimistic scenario tend to use money income or poverty alleviation measures of development, use time series analysis, assume little change in energy intensities, and allow for big boosts to infrastructure spending in developing countries. Researchers arguing that the second, lower-carbon option is feasible tend to use non-monetary measures of wellbeing, use cross-national analysis and assume distinct low-carbon pathways in the East and South.

standards is certain to exceed carbon boundaries. The critical debate continues on whether meeting human needs on a world scale will entail a global lift to much higher emissions or whether it can be achieved via new low-carbon pathways.

Future research should be widened to include a fuller range of need satisfactions for effective participation in different and particular social contexts, though this will be difficult to do on a global scale (but see Rao and Min 2017). To get any closer to the necessary preconditions for decent living, we need to adopt a quite different methodology, that of the 'dual strategy': detailed participative bottom-up discussions and debates informed by expert opinions to draw up lists of necessary items for different types of households and communities, something that can only be done at a national level (see Chapter 7).

But already two conclusions follow from this research. First, the pursuit of basic need satisfaction is in principle satiable and can *always* be met with lower emissions than growth led by untrammelled consumer preferences and expenditure. Second, the more that future paths of development turn out to be high carbon, the more imperative is a global burden-sharing deal, discussed below.

BETWEEN THE SDGS AND 2 °C: CONFLICTS AND SYNERGIES

So far we have treated the two components of sustainable wellbeing – need satisfaction and emissions – as unrelated to each other. But in fact there are numerous causal and more general links between climate change and human wellbeing. This section summarises some conflicts and synergies between the twin pursuit of human development and planetary sustainability.

A major international effort has been made to study the impact of climate change on human wellbeing, once again concentrating on *health* (Corvalan et al. 2005; IPCC 2014b; Smith et al. 2014; Farley et al. 2015; Whitmee et al. 2015). This distinguishes three impacts:

- direct health effects: floods, droughts, heatwaves, storm damage and rising sea levels;
- ecosystem-mediated health effects: infectious disease risks, reduced food yields, mental health, etc.;
- indirect, deferred and displaced health effects: diverse health consequences of livelihood loss, population displacement, conflict and inappropriate adaptation and mitigation; these effects include anxiety, stress and mental ill-health.

The interaction between these variables is complex, but the framework illustrates some of the dilemmas faced in trying to live within Raworth's lifebelt. I discuss first some potential synergies between human wellbeing and climatic stability, second some potential conflicts, and third some of the dilemmas faced in relating the two together.

Potential Synergies

Direct effects: drought, flood and heat

The direct impacts of unmitigated global warming will be more prevalent storms, drought, floods and heatwaves, which will on balance have negative direct impacts on human wellbeing (though some warming in northern latitudes may be beneficial). For example, Watts et al. (2015) document a well-established relationship between extreme, and unexpected, high temperatures and human morbidity and mortality. Heavy labour in hot and humid environments is a particular health risk to workers in the tropics and subtropics. Similarly floods and storms directly threaten life, as demonstrated by Hurricane Katrina in New Orleans and Typhoon Nona in the Philippines.

Another likely impact of unmitigated global warming will be a surge in demand for air conditioning in homes, cars, schools, offices, shops and so on that will feed back on emissions. Demand for cooling in the developing countries of the South and East will grow faster than demand for heating in the North. One recent report concludes that worldwide energy demand for space cooling will overtake that for space heating by 2060.[6]

Indirect effects: air pollution

A major reduction of emissions from burning fossil fuels would reduce the incidence of respiratory diseases. Fine-particulate air pollution is estimated to have been responsible for 7 million additional deaths globally in 2012, mainly due to respiratory and cardiovascular disease. The IPCC's Fifth Assessment Report stresses that the main health co-benefits from climate change mitigation policies come from substituting polluting sources of energy with renewable and cleaner sources, with a considerable effect on the improvement of air quality. For example, since 2000, when the government in the Shanxi province of China promoted several initiatives to reduce coal-burning emissions (including factory shutdowns), the disability-adjusted life years (DALYs) lost in Shanxi decreased by 57 per cent as a consequence (IPCC 2014c).

---

[6] See https://www.theguardian.com/global-development/2016/aug/15/increase-air-condi tioning-accelerate-climate-change.

A shift away from diesel- and petrol-driven transport would also reduce disease from air pollution. Global emissions from motorised transport have doubled since 1970. Moreover policies that encourage active travel (e.g. walking and cycling) produce significant reductions in cardiovascular disease, dementia, diabetes and several cancers, in addition to reduced duration and severity of depressive episodes. It has been estimated that the combination of active travel and lower-emission motor vehicles would give large health benefits across the world, notably from a reduction in the number of years of life lost from coronary heart disease (10–19 per cent in London, 11–25 per cent in Delhi (Woodcock et al. 2009)).

Indirect effects: nutrition
Nearly 800 million people do not receive enough protein and calories for good health, yet a similar and increasing number are now overfed. Food insecurity and under-nutrition, though declining, remains a grave and persistent threat to wellbeing. At the same time there has been an explosion of protein-dense diets in the North and increasingly the East. GHG emissions from ruminant animals such as beef and lamb are at least three times as high per unit of weight as other types of protein and ten times higher than pulses. Meat-eating also requires an extremely high use of water and land per unit of meat. Yet reducing the consumption of red meat also delivers significant health benefits. High meat intake is associated with increased risk of certain cancers and cardiovascular disease. It has been estimated that a 30 per cent reduction in livestock production in the UK would reduce cardiovascular deaths by 15 per cent (Whitmee et al. 2015).

Threats and Conflicts

These are just a few of the co-benefits discussed in the IPCC Report (2014c). But there are also potential conflicts between a greener economy and the pursuit of wellbeing, especially when we move from health to a wider range of Sustainable Development Goals (Cook et al. 2012). Across the world, measures to increase the price of carbon – an absolutely essential component of effective carbon mitigation – will tend to be regressive in the first instance, bearing more harshly on lower-income households and localities (Gough et al. 2011). In the South, the rural poor depend more on directly utilising natural resources, such as forests and pasture; thus policies to price the commons, however environmentally beneficial in the long run, would lower their incomes. Land and food are other areas of intense conflict, as discussed below. Nor is it the case that green sectors are necessarily labour-intensive, thus raising employment for wage labourers. Dercon (2012) concludes that unalloyed green growth may result in the poor being asked to pay the price for sustaining growth while greening the planet: it might engender 'green poverty'. Put starkly, might the double injustice which characterises global environmental relations be converted into a triple injustice, whereby the poor emit less, suffer more *and* bear the brunt of climate mitigation policies?

Are the SDGs Compatible with 2 °C?

Lists of co-benefits and conflicts like the above are useful but not enough. The interactions between alternative mitigation policies, alternative SDG policies, and all the combinations of the two need to be analysed. The two criteria of meeting needs and safeguarding the climate must continually be iterated until they become compatible – or as close to compatible as possible. I consider here just one recent study that attempts to use integrated energy–economy–climate models to compute the implications for certain SDGs of alternative clusters of mitigation pathways consistent with the 2 °C target (von Stechow et al. 2016). In addition to the topics discussed above, this analysis can cover impacts on other SDGs, for example on energy access, full employment, resilient infrastructure, energy security and sustainable production.

Land and food are sites of potentially grave conflict between human wellbeing and climatic stability. Most plans to reduce emissions by 2030, let alone to achieve the longer-term Paris goal of net zero emissions in the second half of the century, rely heavily on *biomass energy with carbon capture and storage* (BECCS). This entails growing trees or plants that absorb CO_2 through photosynthesis, then burning the wood in power stations, capturing the CO_2 from the chimney, liquefying the CO_2, pumping it underground, and storing it for many thousands of years. This whole technology faces major problems: it has never so far worked at scale, there are many technical and economic unknowns, and carbon capture and storage heavily reduces the efficiency of power stations. Above all it will require enormous amounts of land year by year, which is limited and much of it needed for crops. Anderson and Peters (2016) convincingly conclude that BECCS will require a second biosphere and regard the entire strategy as fantasy, yet it is a central component in current plans to keep the climate safe.

By taking land away from food production BECCS threatens SDG Goal 2 (ending hunger and achieving food security for all) and SDG Goal 6 (universal and equitable access to adequate safe and affordable drinking water). It could also threaten biodiversity and place-specific livelihoods. Using more land for bioenergy production will, other things being equal, raise the cost of food, reduce real incomes and thus also obstruct the achievement of SDG Goal 1 (the ending of poverty). Since the share of expenditure devoted to basic food varies from 7 per cent in the US to 70 per cent in Zambia, a major expansion of biomass could worsen both global poverty and global inequality.

> If raw food prices doubled, as happened for rice and corn in the 2007–2008 food crisis, store prices of food in the US would increase by about 18%, in response to which Americans would consume about 1.5% less food. In Zambia, the same doubling of raw food prices would nearly double the price of a food basket, necessarily resulting in a dramatic decline in food consumption. The World Bank estimates that the 2010–2011 surge in food prices drove an additional 44 million people into poverty. (Farley et al. 2015: 247)

One important general finding is that 'optimal climatic pathways' based on least-cost mitigation programmes can conflict with social goals, by ignoring the distribution of costs and benefits across social groups or by incurring unacceptable risks to livelihoods. Many current INDC plans – the foundation of global mitigation effort today – are dangerously inadequate because climate risks are higher than assumed and social co-benefits are lower. There is an urgent need to integrate climate and social development policy at the national level.

One ray of light is that reducing energy *demand*, according to these integrated models, would help reconcile the pursuit of 2 °C and central SDGs.

> Pursuing aggressive energy efficiency improvements across all sectors and *rethinking high-energy lifestyles* therefore seems essential to increase synergies and keep the trade-offs across SDGs manageable in a world that is characterized by multiple constraints ... Future research should ensure that mitigation scenarios are consistent with minimum thresholds of energy demand necessary to satiate basic human needs. (von Stechow et al. 2016: 15, emphasis added)

FROM ECO-EFFICIENCY TO SUSTAINABLE CONSUMPTION

The upshot of the argument so far is that the goal of satisfying universal human needs *could* provide a realistic route to sustainable wellbeing. But there is no simple link between the pursuit of need satisfaction and reducing emissions.[7] Many factors interact when meeting needs is translated

---

[7]   Simon Caney (2012) contends that there is no case for treating the distribution of GHG emissions separately from other issues of global and intergenerational justice, such as development, trade, poverty and health. Poor people do not need or demand 'emissions'; they need 'welfare' or the resources to deliver that. To isolate emissions in this way, he claims,

into *necessary emissions*. They vary between countries and regions, and can be grouped as follows:

a. The structure of the domestic and traded economy. The material throughput on which the delivery of need satisfiers depends will include the entire combination of upstream manufacturing, transport and trading processes. The supply chain will importantly influence the level of eco-efficiency and thus the final carbon footprint of providing necessary goods and services.

b. The level of investment. Calculating necessary emissions should include continuing 'maintenance investment' to repair and replenish manufactured capital goods as they wear out. Developing countries will require a large additional boost to net investment to make good their shortfall in useful capital stock: the emissions entailed must also be taken into account (Rao and Baer 2012).

c. Collective public infrastructure. The provision and maintenance of roads, rail, ports, telecommunications and other infrastructures will emit GHGs and thus affect necessary emissions. This category would include other goods and services mainly provided by states, such as education and healthcare.

d. Geography and climate. As already noted, these factors will affect the energy and emissions required to achieve necessary levels of, say, mobility or warmth.

e. The degree of inequality within societies will also affect aggregate emission levels, as argued in Chapter 3. Unequal, turbo-consumer societies can multiply the range of 'unnecessary' commodities consumed.

f. Cultural influences on preferences. A wide range of cultural factors influence preferences and practices with an impact on emissions. For example, the food-related GHG emissions of the average Brazilian exceed those of the average Indian by 1 tonne, owing largely to the almost fivefold difference in their per capita food-related methane emissions. This highlights the importance of meat consumption, a critical difference between Indian and Brazilian diets (Rao and Baer 2012).

g. The balance between material and non-material satisfiers. Non-material satisfiers, such as our needs for security in childhood, physical security and significant primary relationships, together with the social preconditions for civil and political rights and political participation, and other

---

runs into serious ethical and practical problems, which only an integrated approach can address. This returns Caney to our own normative starting point: 'meeting the basic needs' of all people now and in the future is the fundamental goal. But it convincingly shows that to determine a set of necessary emissions first is to put the cart before the horse.

intangible social factors such as stability and social cohesion all demonstrate the critical importance of non-commodified social relationships to human wellbeing (Caney 2012; Reusser et al. 2013). The extent to which modern economies replace non-material relationships with material commodities will also influence the level of necessary emissions.

This list reveals some of the factors that policy will need to target in order to pursue sustainable and equitable consumption in a climate-constrained world. They help to explain the wide international variations in Figure 4.1 and the existence of countries within Goldemberg's Corner. They clearly show that there is no linear relationship between needs and emissions. Agreement on basic needs does not lead to simple agreement on 'necessary emissions'. The list can be roughly allocated to two categories of climate mitigation:

- *C1. Eco-efficiency of production emissions:* Factors a, b and c affect the efficiency with which energy and emissions are converted into outputs, including outputs of need satisfiers. The extent of the non-commodified economy – g – will also be relevant here.
- *C2. Patterns of consumption:* Factors d, e, f – and also g – will shape the scale and nature of the need satisfiers to be consumed in particular geographical, cultural and socio-economic contexts, and thus the emissions they generate.

This points to a fundamental distinction between two goals: the *eco-efficiency of production* and the *sustainability of consumption*. Each goal motivates different policy interventions. The second points to a policy of recomposing consumption and reducing demand in rich economies. It forms a critical framework for Part II of this book.

FAIR ALLOCATION OF CLIMATE OBLIGATIONS

This then has implications for global equity and the duties, responsibilities and obligations of different nations. Chapter 2 ended with three imperatives for a just approach to climate change: a ceiling or cap on emissions, a fair distribution of the allowable future stock of emissions, and a fair contribution to mitigation and adaptation costs. The criteria of fairness and justice should be applied across three domains: global justice, intergenerational justice and social justice within nations. Together these amount to a strong case for much more equitable 'burden-sharing' in mitigating and adapting to climate change than exists at present.

The Fifth IPCC Report (2014c) distinguishes three rationales for placing equity and burden-sharing at the heart of climate policy: normative, legal (human rights and the earlier UNFCCC commitment to 'common but differentiated responsibility') and political (effective political agreement may hinge on global climate policy being seen as equitable). At the same time it recognises that all such issues are still controversial. How might principles of burden-sharing be applied in practice? How could the costs of an emergency global climate mobilisation be fairly shared between nations and peoples? Most proposals identify two criteria – national *responsibilities* for present and past emissions, and national *capacities* to fund the necessary mitigation efforts.

Allocation of *responsibilities* usually follows the 'polluter pays' principle. A nation's responsibility can be measured by its cumulative emissions of GHGs since a certain date (proposals vary from 1850 to 1998, the year of Kyoto). Clearly the further back the date, the greater the responsibilities of the North. Another issue is whether emissions are calculated on a territorial or consumption basis: territorial emissions are more amenable to verification, but less relevant to wellbeing. However, this is not all: the distribution of responsibility *within* countries should also be judged according to fair or just principles, especially given rising inequality and 'luxury' emissions across the globe. To do this some propose that emissions from people with incomes below an income cut-off point, or development threshold, should be ignored when calculating national responsibilities. Using a crude global threshold of $7500 per capita a year, the Greenhouse Development Rights (GDR) group estimated that high-income nations are responsible for 66 per cent of territorial emissions (WBGU 2009; Baer 2013) (see Table 4.1).

Using a different approach, a study measuring *consumption*-based emissions arrives at a similar figure (Chancel and Piketty 2015). Instead of a money threshold it excludes from its measure of responsibility the bottom

Table 4.1   Greenhouse Development Rights: shares of national obligations up to 2030

Global shares (%)	US	EU	All high-income countries	Rest of world
Responsibility: cumulative emissions	29	21	66	34
Capability: adjusted national income	30	31	82	18
Obligation: average of the above	29	26	74	26

Source: Baer (2013).

90 per cent of the world's population, and concentrates attention on the top 10 per cent of global emitters. This top group also accounts for about 66 per cent of all emissions. North Americans and Europeans predominate; but one-third of high emitters now live outside the North – they are found across all regions of the East and South.

The second normative principle would take into account the *capacity* of a country, its 'ability to fulfil costly obligations without undue or disproportionate sacrifice of welfare'. National capacity is normally equated with GDP per head, but again this can and should be adjusted to exclude all people below a poverty income threshold. On this basis the GDR group considers that nations in the North have 82 per cent of the surplus resources available to fund mitigation and adaptation measures across the world.

Combining the two attributes and weighting them equally yields an overall index of national obligations to combat climate change (see Table 4.1). The conclusion is that the global North bears the lion's share of responsibility for global warming and possesses the economic capacity to do something about it.

The implications are that rich nations cannot rely only on their national carbon reduction targets, discussed in Chapter 3, to make a fair contribution to abating global warming. Owing to lock-in to past high-carbon pathways, there is little chance that their emissions can be reduced fast enough without more radical policies to restrict consumption. Thus they have a greater moral obligation to fund mitigation and adaptation programmes in the developing world. Consider the case of the EU, the best-performing region in the North. The INDC target for 2030 proposed by the EU-28 is for a 40 per cent reduction relative to 1990 emissions. But, if the foregoing arguments are accepted, one study suggests this is dwarfed by the contribution it would need to make to GHG cuts elsewhere if dangerous climate change is to be averted (Athanasiou et al. 2014).

This brings home both the inadequacy of current INDCs and more importantly their conceptual flaw. A global equity framework is needed to reveal the total just contribution of the North and other regions to global climate stability and sustainable development. A substantial part of the North's fair share of mitigation obligations will need to take the form of supporting mitigation abroad, supplying finance and technology to enable countries to develop on a low-carbon pathway.

Yet there are substantial and growing luxury emissions in the global East and South, just as there remain necessary emissions in the global North. The case for a progressive, needs-based allocation of obligations adds to national obligations for climate change a *class* obligation. The Plutocene phenomenon recognises the overriding responsibility of richer people for global warming, wherever they live, but so far no way has been found of incorporating this principle into public policy (Chakravarty et al. 2009).

Critics of the global equity approach claim that it is counterproductive and diversionary. Stern (2015, ch. 9) argues there are no fixed ratios between consumption on the one hand and emissions on the other; indeed policy on climate change is in large measure about altering these coefficients. The list earlier in this chapter spells out seven of these mediating factors, though Stern's focus is on the eco-efficiency of production, not the composition of consumption. Second, Stern suggests that it is politically naïve: to interpret distributive justice in terms of equal allocations is to divert attention from the urgent need to decarbonise the entire world's energy system: 'there is little point in equitable access to a train wreck'. To focus on luxury emissions now is diversionary and bad politics (however 'deeply unattractive rich country intransigence is').

Stern concludes that equity issues remain, but they should take the form proposed by the Indian government at Cancun: 'equitable access to sustainable development'. This comprises a twin strategy of rich countries embarking on a dynamic and attractive transition to the low-carbon economy in their own countries, and supporting that transition in the developing world. However, I would argue that it is difficult to turn this policy into practice without recognising common human needs and the prior moral claims of need satisfiers over preferences for 'luxuries'. Stern neglects the second set of mediating factors listed earlier in the chapter – those stemming from the patterns or levels of consumption.

The IPCC (2014c) report states that 'sustainability is a matter of distributive equity, not of efficiency'. It is critical of orthodox economic valuations that assume constant marginal utility of income and brook no priority for the worst off. But what it lacks is a notion of human need to buttress these claims. A needs-based global equity framework can provide a benchmark for calculating the just contribution of the North and other regions to global climate stability and sustainable development.

The dominant approach today, set out in Chapter 3, is green growth: a strategy to marry the pursuit of growth with rapid decarbonisation. Its policy target is to radically cut the energy and emission costs of production. This is an admirable goal, and the dream of a decarbonised world inspires many green technologies today. But it faces many challenges. The technological transformation required is enormous, and the chances of failing to meet climate and decarbonisation goals are high. In addition, it can make little or no impact on global inequality, which in turn seriously hampers the pursuit of social development goals. Above all, green growth accepts and reinforces the equation of wellbeing with income and prosperity with growth. It excludes from policy debate the role of affluence and surplus consumption.

CONCLUSION

This chapter has returned to human needs and the 'inner boundary' set out in Chapter 1. It asks what would constitute a moral minimum of need satisfaction across today's world and then tries to estimate what claims meeting this minimum would make on the available global carbon space – what 'necessary emissions' it would entail. Meeting needs will always be a lower carbon path than meeting untrammelled consumer preferences financed by ever-growing incomes. But whether it is low enough to protect the needs of future generations will depend on two further factors: the mitigation strategy and the equity framework.

Some mitigation policies like BECCS, discussed above, could threaten the very systems on which current wellbeing depends. Most crucially, all existing strategies ignore the role of consumption levels and patterns in the affluent world. Yet the more that future paths of social development turn out to be high carbon, the more imperative is a fair system of global burden-sharing between North and South. Sustainability and distribution are intimately connected. My conclusion is that equity, redistribution and prioritising human needs, far from being diversions from the basic task of decarbonising the economy, are critical climate policies. What this implies for rich countries is the subject of Part II of the book.

PART II

Towards eco-social policy in the rich world

5. From welfare states to climate mitigation states?

Part II of this book is concerned with the rich world of the 'North': that is, the advanced capitalist countries that combine membership of the OECD with Annex I of the United Nations Framework Convention on Climate Change (UNFCCC).

These countries are responsible at present for a major share of the cumulative consumption emissions of GHGs causing global warming. Their share of global wealth and income means that they have the greatest capacity to fund global climate mitigation and adaptation measures. As the core capitalist economies, notably the G7 countries and especially the US, they play a powerful leading role in global economic governance. And they possess in the social dimension a characteristic institutional form of social policy, namely *welfare states*. The interrelation of these factors gives the political economy of the North a particular complexion and dynamic, which is the main focus of this chapter.

State responsibility for social protection, universal education, health services and more has been a notable feature of these countries since the Second World War, so much so that it has become commonplace to speak of welfare state capitalism or democratic welfare capitalism. Concern with environmental conditions is a more recent development dating from the 1970s, and state concern with climate change more recent still, dating from the 1990s. These new national responsibilities and interventions have been layered on top of existing economic management functions and forms of capitalism, so that social and climate policies are partly shaped by them. Despite the shift towards globalisation, nation states remain the central actors in today's world, and that is especially true of a small set of powerful states, plus one unique multi-level actor, the EU – a point well illustrated by the emergence of national plans for GHG reductions up to the 2015 Paris climate summit. It is this political economy terrain that has generated major climate change, and it is here that major decisions about how to address it will be decided.

In the twenty-first century we must find new ways of linking the old 'social question' of the nineteenth and twentieth centuries with the new climate challenge of planetary sustainability. There are already growing calls to integrate sustainability, efficiency and justice goals within the world of democratic welfare states. Just as the post-war 'social settlement' can be interpreted as an attempt to reconcile the goals of efficiency and social justice by combining capitalist markets and ownership with full employment and a 'welfare state', so there are now calls for a 'new social settlement' to combine 'welfare capitalism' with environmental sustainability. These calls have come from a wide range of political and intellectual positions, for example in the UK from Dieter Helm at Oxford University and the New Economics Foundation in London (Helm 2011; Coote 2015).

The chapter begins by summarising – in the context of social policy – the main ways in which climate change is likely to impact on the North, directly and indirectly. There are severe risks that will challenge the capacity of these countries to protect their populations. But in addition they need to undertake a massive and urgent programme of decarbonisation to cut emissions in both production and consumption, which presents an even greater set of challenges. The rest of the chapter discusses the prospects.

It describes and analyses the emergence of welfare states in the twentieth century and the subsequent beginnings of climate policy at the end of that century and into the new millennium. It notes a range of factors that have shaped the two epochal shifts in state intervention, and considers the parallels and differences between them. It then explores cross-national variations among the Northern states that shape distinctive welfare regimes and variations in climate strategies, offering some analysis to explain these patterns and their significance. The chapter concludes by introducing the remaining chapters, which analyse, in turn, three distinctive approaches to climate mitigation and how they interact with key social policy goals.

SOCIAL POLICY AND THE IMPACTS OF CLIMATE CHANGE

Social policy is frequently defined as the public management of social risks (Esping-Andersen 1999). As such it is certainly not a new policy area, having existed for several millennia in major ethico-religious traditions and empires.[1] In addition, a 'core' economy or 'social' economy has always existed alongside market relations: caring for and socialising children within households, building and maintaining communities, creating shared meanings and social cooperation. These activities are always essentially unwaged and are still mainly undertaken by women. Without

---

[1]   For example, charitable institutions, food buffer stocks, quarantines, flood control and so on developed long before modern times (Gough and Therborn 2010).

these uncommodified activities, as feminist analysts stress, markets and waged work could not continue (Elson 1988; Mellor 1997; Barry 2012; Fraser 2014).

As market economies became more self-regulating with laissez-faire capitalism through the nineteenth century, so a distinct realm of *social policy* began to take shape. Social programmes, such as public health, elementary education, factory legislation and social insurance, were introduced in response to new risks and insecurities in the labour market, pressures from trade unions, social democratic and communist parties and movements, fears of social unrest among governing elites, and the rise of national discourses around the 'social question'. Following the Second World War, these programmes became institutionalised in a variety of *welfare states*.

Climate change is not only an overwhelming new risk – global, long-term, persistent and cumulative – but also a profound uncertainty clouding all our futures. It confronts contemporary social policy with a qualitatively new agenda. So far, there has been little discussion or research on the linkages. Some new scholarship is now slowly changing that picture (see for example Fitzpatrick 2011, 2014a, 2014b; Schaffrin 2014; Koch and Mont 2016). Tellingly, however, there is no reference to 'social policy' in the index of the 1400-page *Mitigation of Climate Change: Contribution of Working Group III to the Fifth Assessment Report* (IPCC 2014c).

So what are the risks presented by climate change in the rich North? They can be divided into four categories (Gough et al. 2008a):

- impacts of climate change itself, distinguishing:
  - direct impacts in the North,
  - indirect results in the North of impacts elsewhere;

- the impacts (direct and indirect) of climate change *policies*, distinguishing:
  - adaptation policies,
  - mitigation policies.

These are surveyed briefly below, with the main focus on the first three, as mitigation is dealt with in detail in the next chapter.

Direct Impacts of Climate Change

The direct impacts of climate change on habitats will hit tropical and subtropical areas harder and earlier, but this does not mean the northern, richer world will be unaffected. This applies to every geographical zone encompassing the OECD world including Europe, North America, Australasia, and Japan and Korea. To simplify the picture I will concentrate here on Europe. Here there are predictions of more frequent and intense heatwaves, increasing rainfall and much heavier winter rains in northern and parts of central Europe, with less rain and more extensive drought areas in the Mediterranean zone and parts of central Europe (IPCC 2013).

One consistent conclusion is of a north–south gradient within Europe. The climate risks are greater the further south you go, while parts of the north could enjoy certain medium-term benefits – a longer growing season and increased yields in agriculture, increased forest productivity, a northward shift in fish stocks and reduced demand for space heating. New threats include rising flood and storm damage and summer heat stress. Southern Europe will experience rising heat and water stress that will trigger new risks to water availability and present benefits stemming from ecosystem services. It will also suffer a range of threats to wellbeing including the impacts of excessive heat on population health. Climate change will have more widespread negative impacts on economic activity in southern Europe, for example on tourism and agriculture. One exception to the north–south gradient is the Arctic, which is likely to experience dramatic temperature rises with direct consequences for the economies and cultures of indigenous communities (IPCC 2014b; Johansson et al. 2016).

Alongside the accumulating impacts of climate change we must consider its distributive effects. Some groups and populations will be more vulnerable to climate change than others, depending on a range of socioeconomic capacities and resources (see Chapter 1). In the developed world, lower-income households are more likely to live in higher-risk areas, marginal lands and floodplains; they have fewer resources to cope and have much lower rates of insurance cover; they may also suffer from poorer health, which in turn undermines their resilience (Walker and Burningham 2011). In critical moments of climate-related disaster this social dimension is brutally revealed, as with the impact of Hurricane Katrina in New Orleans in 2005. The gravity of a 'natural' disaster is never purely natural. Social protection policies, safety nets and public services can offset some of the effects.

Indirect Global Impacts and Climate-Induced Migration

There are numerous indirect ways in which the heating of the planet could and will impact on countries in the global North. The UK Foresight Report catalogued a vast list, including resource scarcity, epidemics, degraded coastal infrastructure impeding shipping, disruption of vital oil and gas supplies, insecurity of food supplies with rising and more volatile prices, disruption of international economic networks and chains, growing restrictions on free trading which may harm the City of London, slowing global economic growth, collapse of weak states, and growing international tensions weakening global governance (Government Office for Science 2011). Here, I concentrate on one potential impact: distress migration from South to North, in particular to Europe (though most climate-related migration is internal to nations).

There are four possible pathways through which climate change could stimulate migration (Martin 2010): the intensification of natural disasters; increased warming and drought that affects agricultural production, employment and access to clean water; sea level rise, which makes coastal areas and some island states increasingly uninhabitable; and competition over natural resources, which leads to conflict and displacement of inhabitants.

The causal chains and links between climate change and migration are complex, but literature reviews and historical evidence strongly suggest that changes in climatic conditions have already been a contributory factor in human migration. Climate-induced migration, like other forms of migration, sits along a continuum from those seeking a better life to those fleeing to save their lives. Climate change can threaten the satisfaction of many people's basic needs and thus 'climate refugees' should be covered by existing human rights declarations and accords. However, there is currently no category in the United Nations High Commission for Refugees classification system for environmental refugees. Some slow progress has been made in recognising claims for compensation, but this does not yet extend to states' responsibilities to receive climate migrants.[2]

European experience in 2014–15 following large inflows of political refugees and asylum seekers from Syria, Afghanistan, Eritrea and elsewhere does not bode well. The burden has fallen on front-line states, notably Greece, Italy and Hungary, and certain destination states, notably Germany and Sweden. EU proposals to share this burden are currently stalled. This illustrates the difficulties facing a Europe with divided levels

---

[2] In 2013 parties to the UNFCCC agreed to establish the Warsaw International Mechanism for Loss and Damage Associated with Climate Change Impacts, to support countries that are particularly vulnerable to the impacts of unavoidable climate change, extending to population displacement and climate-induced migration (Eckersley 2015). China and the G77 have argued that 'loss and damage' goes beyond what can be reduced by successful climate adaptation, and should therefore form a third pillar of the UNFCCC in addition to mitigation and adaptation. However, this was successfully opposed by the US at the 2015 Paris COP.

of governance. It would seem inevitable that climate migration pressures from Africa and elsewhere in the global South will build up in the Mediterranean region. Latest research shows that adaptive capacity is stronger in north-western Europe than in the Mediterranean area, where climate stress will be greater (Crouch 2015; Johansson et al. 2016). This suggests an emerging triple burden for southern Europe: greater climate stress, weaker adaptive capacities and greater migration pressure if the predictions about climate refugees prove correct.

A more unified EU-level system of governance would in theory be better able to handle some of the intra-European implications. However, in continental-size nation states like the US and Australia, where there are more unified responses, these vary widely – from hard-line borders around Australia to more prevalent access (though increasingly contested) to the US.

New patterns of migration highlight an inherent tension within welfare states: entitlements delivered to citizens are seldom extended to illegal residents or to other non-citizens and 'denizens', with rising pressures from migration leading to 'fortress' welfare systems (Christoff 2013).[3] Climate change will test the ability of national welfare states to internationalise and recognise collective responsibility for meeting the basic needs of people elsewhere in the world. Existing welfare states are national responses to need and will find it extremely difficult to expand the boundaries of the moral community to include the natural environment and the needs of distant strangers (Eckersley 2004).

Impacts of Adaptation Policies

Adaptation measures are intended to reduce the vulnerability of natural and human systems to actual or expected climate change effects (see Chapter 1). Rich countries will need to take necessary measures: strengthening critical infrastructure, such as flood defences to protect against storm surges, making buildings more resilient to weather extremes; securing long-term provision of water, food, power and transport, such as extra reservoir capacity; and providing emergency assistance and relief (Christoff 2013).

Growing demands for climate adaptation will exert more pressure on strained fiscal and welfare systems. Thus far, political scientists have found

---

[3]   After the Second World War, Germany dealt with millions of displaced people. France later admitted large numbers of Algerians. These may provide some lessons for dealing with climate refugees, but they mainly dealt with displaced 'nationals' (note from Stephan Leibfried).

that welfare policies prevail over environmental policies. This is because direct human impacts from social ills such as poverty and disease usually bite harder and/or more immediately than do indirect effects of environmental deterioration, and because welfare systems have nurtured interest coalitions in their support. But as concerns about climate change grow they could increasingly preoccupy decision-makers, pre-empt available fiscal resources and thus undermine established social policy objectives (Gough and Meadowcroft 2011).

The IPCC concludes that Europe enjoys relatively high adaptive capacity (IPCC 2014b). Public welfare systems themselves constitute part of climate adaptation systems. Yet the spatial scale of what remain essentially *national* welfare systems is problematic. The US, Canada and Australia cover more than one IPCC climatic region, which suggests more flexible opportunities to combine climate adaptation and social protection. Europe spreads over three climatic regions but comprises many nation states. It provides strong arguments for a regional EU strategy that combines both welfare systems and climate adaptation.

Impacts of Mitigation Policies

Climate mitigation policy refers to all human interventions to reduce the sources, or enhance the 'sinks', of greenhouse gases. It covers everything from preventing the future mining of coal and oil reserves to labelling refrigerators, from supporting solar power to insulating lofts in houses, from reforestation to taxing gasoline. But all can impact on human wellbeing in the affluent world, both monetary and non-monetary wellbeing, both in aggregate terms and in terms of the distribution of costs and benefits. The ultra-fast climate mitigation that is now needed, especially in the global North, will upset taken-for-granted economic and social practices in every domain of life. It will impact hard on some sectors, regions and people and prove beneficial to others. It will test governance and democracy. For these reasons the entire next chapter is devoted to considering certain climate mitigation strategies and their social consequences in the affluent world.

In conclusion climate change brings with it new configurations of risk, injury and inequity for social policy. The scale, range, combination and timing of problems are uncertain, but the effects are potentially huge. Climate-related polices will create winners and losers. Social policy frameworks will require a great deal of adjustment simply to maintain levels of equity and social justice – let alone to improve them. On the other hand, social policy could play an important productive role investing in human, social and community resources, thus helping to mobilise collective policies to adapt to and mitigate climate change.

WELFARE STATES AND THE CHALLENGE OF CLIMATE CHANGE

Welfare states transfer the allocation of some goods and services from the market to political determination, often in the form of 'social rights of citizenship' (Marshall 1950). Thus all rich OECD countries have extensive social security systems covering old age, disability, sickness, unemployment and other contingencies, plus comprehensive public education systems. Most, excepting notably the US, also have universal healthcare entitlements and child allowances and other family programmes (the future of Obama's Affordable Care Act of 2010 remains uncertain). Today, in the long-standing OECD member states, average social expenditure, excluding education, accounts for around 23 per cent of GDP, funded from significant tax and social contributions. This expansion began in the first three decades after the Second World War, but it has continued since then – growing by another 5 per cent of GDP in the quarter-century since 1980. After the Second World War the social policy domain developed alongside Keynesian macro-economic policy aimed at managing total demand and maintaining something close to 'full employment'. Thus welfare states include major interventions in labour markets as well as redistribution.

If we are to understand their actual and potential capacity for meeting the climate challenge, we need to map out and explain their post-war development.

A Brief History of Welfare States

Recognisable welfare states emerged from the end of the Second World War, though in many European countries the first post-war years were scenes of largely unrelieved hardship and displacement of peoples. But once under way the years up to the mid-1970s were exceptionally favourable for welfare states for several reasons. Capital was relatively immobile in the initial period of the post-war settlement, so there was considerable room for redistribution. As trade protection was gradually removed with lower tariffs, social protection took its place. The experience of war and depression paved the way for the emergence of a Keynesian consensus promoting high levels of employment, high tax and expenditure levels and a dominant ideology favouring government management of demand and the business cycle in capitalist economies. Distributional conflicts were mitigated by a comparatively symmetric balance of power between the interest organisations of labour and capital and by relatively high rates of economic growth. Moreover the Cold War and competition with the Soviet Union encouraged social inclusion. Under these circumstances, social benefits were everywhere significantly raised, existing programmes were extended to cover new groups of beneficiaries, and entirely new schemes were adopted. Social outcomes improved, including a decrease in inequality and poverty, the partial 'de-commodification' of labour, the assurance of social rights and improved macro-economic performance.

Things began to change in the late 1970s. We can discern across the OECD a *counter-counter*-movement against social democratic ideas of the 'Keynesian welfare state', often labelled *neoliberalism*. Its defining ideas included a belief in the superiority of markets and a denigration of government and collective action. Its defining characteristics included a new international division of labour, the global spread of production networks, trade and financial flows, the dominance of finance, rising profit shares and widening inequalities within countries (Newell and Paterson 2010; Koch 2012; Stiglitz 2013; Piketty 2014). A post-industrial system began to emerge in the OECD world, where new social risks were layered on to older risk patterns. Flexible, temporary and atypical forms of work generated new forms of insecurity. Large numbers of women entered the labour force with positive impacts on their life chances and household incomes. Major shifts in family structures took place with rising divorce and more recombined and single-parent families. A growing service economy with lower productivity, combined with economic globalisation, imposed tighter constraints on public revenues. Business and finance developed more direct influence over policy-making, while the power of trade unions declined, especially in the Anglosphere.

In general, welfare states proved more immune to expenditure retrenchment than other policy areas such as infrastructure and economic affairs. Even in the US, despite relentless ideological and fiscal pressures, core programmes such as Social Security and Medicare proved remarkably durable. Across the OECD, social policy continued to redistribute incomes from richer to poorer: a comparison of post-redistribution household incomes with original market incomes in 2012 showed that poverty levels were reduced by about two-thirds and inequality across them reduced by around two-fifths (Ferragina et al. 2015). Public provision of education and health services at low or zero costs to the user also had highly redistributive effects.

Against this, however, spending did not keep up with rising social needs, which were driven by demography, family change and socio-economic shifts, coupled with Baumol's 'cost disease' (low productivity gains in human services where workers' earnings tend to rise in line with the private sector). Life expectancy continued to rise, while in many European countries fertility stayed below replacement rates, so that pensions and medical care absorbed a growing proportion of public funds. Overall, service levels and social rights diminished somewhat between the early 1980s and the late 2000s (Kühner 2015). It is important to note that these demographic and other trends have imposed costs on all pension and health systems, whether public or private; the difference is that the costs can be shared more fairly in welfare states.

Since 1980 welfare states have coexisted with a historic rise in inequality (described in Chapter 3). Across the OECD, but especially in the US and the UK, the distribution of income has favoured the rich, whatever the measure used. In the US the top 10 per cent acquired almost the entirety of economic growth in the early 2000s, while the real incomes of the 90 per cent stagnated. According to Tomas Piketty (2014) the systemic tendency for the share of capital and inequality of incomes to rise under capitalism reasserted itself after around 1980, following several decades when countervailing forces prevailed.

The world financial crisis of 2008 transformed the fiscal environment for social policies. According to many economists the crisis was itself a product of soaring inequalities generated since 1980 by the new phase of financialised capitalism, which fuelled cheap credit and consumer debt and unbalanced economies. The subsequent bank bailouts and the ensuing recession transformed a private sector banking crisis into a fiscal and sovereign debt crisis. A prolonged recession in many OECD countries reduced tax revenues and drove up public sector deficits and the stock of government debt. In the event it was risks facing the financial sector that were socialised, not those facing workers and families. Real incomes fell and in some countries, such as the UK, they took years to recover (Gough 2011b; Stiglitz 2013; Kumhof et al. 2015).

'Austerity' has been the dominant policy response: aiming to get borrowing and debt down by cutting government spending and raising taxes – though the ratio is heavily biased towards cuts. This has exacerbated the decline in real incomes and consumption spending, which has worsened the economic downturn and, in many countries, exacerbated levels of public and private debt. In the UK the crisis has been used to radically refashion the post-war welfare settlement and bring it closer to the American model (Taylor-Gooby and Stoker 2011; Taylor-Gooby 2013; Streeck 2014).

Debates over the economic impact of welfare states on market economies have fluctuated between 'compatibility' and 'incompatibility' positions. Neoclassical economists and neoliberals argue that welfare states undermine economic performance and national competitiveness: Okun's (1975) overused metaphor of the 'leaky bucket' argues that total income is reduced when money is carried from the rich to the poor. Yet other economists such as Tony Atkinson (2015) maintain that social protection, labour market security and social provision are compatible with and augment productivity and competitiveness. Economic sociologists and political economists also show that the relationship between welfare and competitiveness is contingent on the institutional context, both within a country and in terms of its position in the global economy (Gough 1996, 2000, ch. 8; Smelser and Swedberg 2010). Stiglitz has demonstrated that growing inequality is associated with less efficient economies, flourishing rent-seeking and non-ethical behaviour, a loss of social trust and erosion of the rule of law, lower public investment and 'the evisceration of democracy' (Stiglitz 2013, chs 3–5; cf. Turner 2012, ch. 1). The IMF (2014) now finds that 'redistribution appears generally benign in terms of its impact on growth'.

I would take a more nuanced position. Welfare states need have no detrimental effect on the capitalist economies in which they sit, as long as they meet three certain conditions: their expenditures should serve to reproduce the socio-economic system as a whole; they should be financed mainly from taxes levied on earnings not on wealth or income from capital; and public services should increasingly procure from, and contract out to, the private sector. But of course meeting these stipulations will threaten the more fundamental purpose of welfare states, which is to enhance human wellbeing. The first condition prioritises 'productive social policy' (explained below) over the claims of the care and wellbeing of 'non-productive' groups. The second condition limits any sustained move towards more equality and prioritising human needs. And the third means commodifying the delivery of numerous educational, health, personal and care services, subsuming their social value within a crude monetary calculus, and giving rise to extensive market failures and losses in welfare (Gough 1979, 1996, 2000, ch. 8).

There is a continuing *tension* between social need and the market economy, and between the claims of citizenship and the drive towards profit. The balance between different models of provision and wellbeing outcomes will reflect the dominant ideas, interests and institutions of the time. The British sociologist T.H. Marshall declared seven decades ago that, 'in the twentieth century, citizenship and the capitalist class system have been at war' (1950: 84). Put another way, all existing welfare states are 'contradictory' (Gough 1979). In the twenty-first century the struggle has become increasing one-sided. The likely scenario at the time of writing is that post-2008 the OECD world will face high or rising inequalities *and* significant reversals of social spending and welfare state provision *and* slower rates of economic growth. If present trends in social policy continue, climate pressures and the demands of mitigation will land on a system of challenged or degraded welfare states.

It is difficult to distil from all this movement, variation and contestation any core definition of social policy or to map its scope (Béland and Mahon 2016). Nevertheless it is possible to identify three broadly progressive goals implicit in welfare states. These matter because, when it comes to addressing the climate challenge, we can look to them to play a role in pursuing sustainable justice and wellbeing. The first goal is to redistribute income: establishing an income floor, modifying market incomes including wages and labour conditions, reducing inequality and redistributing according to need. The second is to promote social consumption: collectively providing vital need satisfiers and discouraging harmful behaviour. The third is social investment: building human and social capabilities. Table 5.1 sets out these goals to provide a framework for subsequent analysis. I will add climate-related data in the following chapters.

Table 5.1   The generic goals and activities of social policy relevant to climate mitigation

	Welfare goals	Typical existing social programmes
Redistribution	Guarantee a minimum income floor. Reduce inequality.	Social protection programmes: public money transfers to individuals and households, including social insurance, universal, income-related and tax benefit programmes. Taxation: in itself, and via 'fiscal welfare', a major redistributive instrument.
Social consumption	Collectively provide vital need satisfiers. Regulate and prevent harmful consumption.	Social services in kind: free or subsidised access to education, health services, social care, housing, etc. Regulation and prevention: policies to modify social behaviour and patterns of consumption.
Social investment	Develop human and social capabilities.	Education, training, equal opportunities. Employment policies, ranging from 'full employment' policy, labour market regulation and wages policies to 'activation' and working-time reduction programmes.

Explaining the emergence of climate policy

Against this background of established welfare states, we can discuss if, when and how far 'environmental states' or 'eco-states' have now emerged, and note parallels and differences between the two. Both welfare and environmental policy interventions can be conceived as political responses to long-term societal change related to industrialisation, urbanisation and democratisation. Both have been called into being to wrestle with issues that cannot adequately be addressed by markets and voluntary action. Both create new patterns of economic interaction (through regulation, fiscal transfers and so on), while operating within significant economic and political constraints. Both point to new forms of political economy and perhaps governance (Lundqvist 2001; Dryzek et al. 2003; Eckersley 2004; Meadowcroft 2005, 2012).

In the 1970s governments began to develop responsibilities and programmes to manage environmental problems, so that an independent realm of law, policies, administration and regulation emerged. In the late 1980s–90s, a second phase began to strengthen linkages between this apparatus and broader state concerns with the economy, security and welfare. A study of 25 environmental policy areas across 24 countries in the global North and East finds a clear, strong and parallel rise in policy interventions from 1970 to 2005 (Sommerer and Lim 2016). The North has led, but all the emerging economies surveyed in the East have kept pace; there is no uniform evidence that regulatory measures have stalled to favour economic competitiveness. Of course, this trend does not imply that the policies to tackle environmental problems have been adequate, effective or timely.

Climate adaptation programmes form part of a long-standing environmental agenda that includes, for example, measures to tackle pollution and protect natural resources. Climate mitigation is of a different order. By its very nature it is a trans-boundary and global challenge, which means that national programmes are likely to take place within a framework of global or regional discourses, declarations, treaties and so on (Eckersley 2016; Sommerer and Lim 2016).

What are the factors that influence and shape the willingness and capacity of states to deal with social and climate risks and in particular to undertake serious climate mitigation? Are the same factors at work in the domains of social and climate policy, or are they different? Studies of specific national *climate* policies and achievements are still relatively scarce (Christoff and Eckersley 2011; cf. chapters in Duit 2014; Fankhauser 2014; Koch and Fritz 2014).

As a way of ordering this research I utilise and apply an earlier framework (Gough 2008, 2016). This posits five drivers of social policy development:

*Table 5.2   Selective major drivers of welfare states and climate mitigation
             states post-1980*

Driver	Welfare state	Climate mitigation state
Industrialisation and structural problem pressures	Growth plus deindustrialisation. New social risks. Demographic trends. Rising inequality.	Consumption growth plus carboniferous capitalism. Ever-tightening carbon and GHG constraints.
Interests	Balance of class interests. Rising structural power of financial capital.	Balance of 'brown' and 'green' business interests. Green movements and their political expression.
Institutions	Pluralist versus corporatist interest representation. Centralised administrative state. Prior welfare regime.	Pluralist versus corporatist interest representation. Extent of devolved powers to lower-tier authorities. Prior welfare regime.
Ideas	Neoliberalism versus social citizenship. 'Social investment' paradigm. Degree of political polarisation.	Neoliberalism versus 'ecological modernisation' and 'green growth'. Degree of political polarisation.
International influences	Economic globalisation and its management.	Economic globalisation and its management. International linkages. Policy diffusion.

Source: Gough (2016: 36).

the 'five Is' of industrialisation (and other problem pressures), interests, institutions, ideas/ideologies and international influences. 'Drivers' here refer to all factors that together condition and mediate the development of the social and environmental interventions of states. They include factors that lead to their different configurations in different national contexts. This framework is not itself a theory but a heuristic which embraces all the major theoretical currents in comparative welfare state research.

Table 5.2 summarises the findings (Gough 2016). First, structural *problem pressures*, stemming from globalisation, financialisation, deindustrialisation and ongoing but slower growth, have been critical background drivers in both domains but in contradictory ways. Continuing growth of production and consumption in the OECD has precipitated new environmental problems and driven dangerous levels of global warming.

Meanwhile for welfare states the slide in growth rates during and after the 1970s contributed to various forms of retrenchment over the next three decades. A different problem pressure that might be expected to affect climate policy is the vulnerability of countries to the impacts of climate change. Yet it appears to be a poor predictor: for example, Australia is vulnerable on many fronts but lags behind in climate policy.

Working-class *interests* have been weakened in the social domain and feature little in the environment domain. The impact of their replacement by environmental and green movements has yet to be evaluated. Business interests have a renewed salience in both domains but in different ways. Welfare states have been constrained by the growing structural power of business and finance as a result of globalisation and technical change; however, the scope and impact of business on environmental states are affected by the relative power of 'green' and 'brown' business interests.

The impact of political and economic *institutions* has been more similar across the two domains. Factors favouring both welfare and environmental states appear to be both political (proportional representation with substantial green parties versus first-past-the-post electoral systems, etc.) and economic (long-term versus short-term financial structures, corporatist versus liberalised corporate and labour market institutions). On the other hand, the recognition that localities have a critical role to play in environmental innovation sits uneasily alongside the role of centralised administrative agencies in welfare states.

The *ideas* and ideologies shaping welfare and environmental states have been crucially affected by their timing: welfare states flourished in the decades of Keynesianism and social reformism. Despite the earlier flowering of green awareness, climate states have developed in the neoliberal decades which have denigrated and sought to unravel state capacities and regulation: the tragedy of 'Bad Timing' according to Naomi Klein. This has also undermined and reversed some features of social policy in different countries, but the persistence of political constituencies and path dependencies has offered some protection. The presence or absence of polarising political discourse is relevant in both domains – 'anti-welfare' versus 'social investment'; climate denial versus green growth – and explains persistent differences, notably between the US and other advanced economies.

International influences have had very different impacts on welfare states and environmental states. Globalisation, economic openness and competitive pressures have acted as a constraint on welfare states. In the environmental domain it has worked both ways. On the one hand it has accelerated economic development that is spreading environmental dislocation to new zones of the world. On the other hand, international linkages appear to facilitate the adoption of environmental and climate change policies. Climate change policies have also been driven by cross-national policy learning and diffusion, particularly evident within the EU over this period.

VARIETIES OF WELFARE REGIME AND CLIMATE MITIGATION

It is clear that comparative research reveals not just common trends but also consistent national variations across the OECD. These variations are important in understanding how social and climate policies can collaborate or conflict in the future. Exploring the differences between national regimes – of welfare, and of environmental and climate mitigation policy – and how they relate to each other helps to identify characteristics that are more or less conducive to addressing the climate challenge. (These questions are returned to in Chapter 9.)

Different *welfare state regimes* had become apparent by the 1970s, varying according to their policy structures and their social outcomes. Esping-Andersen (1990) identified three: 1) a social democratic or Nordic model with high levels of labour de-commodification and cross-class solidarity, resulting in a system of generous universal benefits and a strong state role; 2) a liberal or Anglophone model, with typically low levels of de-commodification, more targeted welfare benefits, and a strong preference for private welfare spending; and 3) a conservative/continental model manifesting a moderate to high degree of de-commodification, a narrower sphere of solidarity related to occupational status, and a commitment to subsidiarity and the preservation of traditional family structures typical of continental Europe. Other regimes have since been proposed, including one based on Southern welfare capitalism in Italy, Spain, Portugal and Greece. There has been considerable empirical support for this regime-based approach since then. In analysing the impact of climate change, we must therefore distinguish impacts common to all welfare states from those which differ according to the type of welfare regime.

Despite the neoliberal reaction of the 1980s and the impact of the later financial crisis, the institutional differences between liberal, social democratic and conservative-corporatist regimes have persisted and in some respects intensified, with marked impacts on national wellbeing. For example inequality, as measured by the Gini coefficient, varies from a high 0.38 in the US to 0.23 in Denmark and Sweden, and the poverty rate in the US remains over three times higher than in the Nordic countries. A wealth of evidence shows that northern Europe remains more egalitarian and experiences less insecurity and inequality than the Anglosphere and southern Europe (Goodin et al. 1999; Hay and Wincott 2012; Crouch 2015). In other words, domestic institutional features continue to play an important role in mediating globalisation (Pierson 2001; Swank 2002).

The Nordic countries have also pioneered the idea of 'social investment' or a *productive* welfare state, later adopted by the EU Lisbon Treaty (Hemerijck 2012). The strategy stresses the role of social policy in supporting productivity, and prioritises education, early years learning, active labour market policies and family-friendly policies to enable more parents and carers to undertake paid work. It links economic and social progress and gives a prominent role to social capital. Much hope is now invested in social investment across the OECD, though the most recent research finds meagre evidence of any real progress outside Scandinavia: 'the general picture is that the social investment glass is almost empty' (Morel 2012).

Turning to national differences in *climate* policy, Robyn Eckersley has summarised the scholarly literature as follows (with some implicit links to the model in Table 5.2). The strongest prediction of a national climate 'leader' would be a developed state with the following characteristics: absence of deep polarisation on climate science and climate action; weak dependence on fossil fuels and a diversified economy and industry associations; a relatively stable population or low population growth rate and therefore stable or falling energy consumption; a social market philosophy; a strong tradition of scientific research; a precautionary risk culture; a strong and diversified media; corporatist/consensual political institutions with a proportional representation electoral system; minimal political veto points (e.g. unitary state, unicameral parliament); and an internationalist or cosmopolitan conception and narrative of the state's role in the world.

Conversely, a national climate 'laggard' would possess the very opposite of these features: deep political polarisation on climate change; heavy dependence on fossil fuels with strong regional dependencies and minimal economic diversification; strong fossil fuel lobby groups; a neoliberal economy; a growing population and therefore rising energy consumption; a majoritarian and adversarial political system; multiple political veto points (federal system, bicameral parliament with strong upper chamber and/or presidential system); a weak tradition of scientific research; a risk-taking culture; a concentrated commercial media; and a self-regarding or nationalist foreign policy tradition.

Of course, no such state of either type exists, but certain countries have more of the first set of features than others, such as Sweden, Denmark and Germany. And other countries, notably the US, Australia and Canada, share many, though not all, of the latter features. This analysis helps to explain why climate denialism is most prominent in the Anglosphere, muted in Western Europe and absent in most of the developing world.[4]

The one obvious exception to this pattern is the UK, which exhibits few predisposing factors but which currently figures as a climate change 'leader' (discussed in Chapter 6). I would hypothesise that this is due to three interrelated features of the UK political economy. First, advanced deindustrialisation in the UK means there are few business or trade union interests very strongly opposed to carbon mitigation; second, the huge accumulated trade deficits 'offshore' a quarter of UK emissions; and, third, carbon trading and other climate-related financial activities offer growing opportunities for profitable trading in the City of London. We return to these questions in Chapter 9.

CONCLUSION: FROM THE SOCIAL QUESTION TO THE CLIMATE CHALLENGE

This chapter began by mapping some of the likely impacts of upcoming climate change on the affluent welfare capitalist nations. These nations have begun to implement variable carbon mitigation strategies (a full analysis of these is postponed to the next chapter), and the chapter sets out a political economy framework to understand these policy innovations and to relate them to social policy developments. Such a survey reveals both common trends and significant national and regional variations. The chapter provides a framework for thinking about the relationship between the essential need to decarbonise our way of life in affluent societies and the need to ensure that such mitigation is broadly equitable and just, despite growing insecurity and inequality within the economy.

In the following chapters, I explore three logically distinct (but not mutually exclusive) approaches to meeting the climate challenge in affluent countries. The first two I have identified in Part I: C1 – raising the eco-efficiency of production; and C2 – recomposing consumption. Because my focus is now on the North, with its special responsibilities for decarbonisation, a third approach is added: C3 – cutting total demand and ending our commitment to economic growth.

Chapter 6 describes current programmes to decarbonise *production*, or in other words to improve eco-efficiency, in the UK and other rich countries. It examines their social and distributive consequences, and considers some eco-social policies to marry equity and decarbonising goals.

---

[4] Robyn Eckersley, Climate leadership before and after the Paris agreement. Public lecture at the LSE, 17 November 2016.

Chapter 7 switches attention to *consumption* in the global North, a domain of potential overlap between climate and social policy. It distinguishes necessary from surplus consumption and discovers an awkward dilemma: many necessities are higher carbon than many 'luxuries', suggesting that traditional redistributive social policy will raise, not lower, emissions. To overcome this dilemma the chapter proposes further eco-social policies to recompose consumption in an equitable way and situates these within a comprehensive strategy for prevention.

Chapter 8 takes off from the argument that even a combination of radical eco-efficiency and recomposing consumption will not cut emissions fast enough to avoid dangerous global warming if economic growth continues in the rich world. It investigates arguments for degrowth and considers the implications for present-day welfare states. It advocates further eco-social policies to restrain consumption absolutely, including shorter working hours.

Chapter 9 concludes by arguing that these three strategies could form a transitional path from the present to a future of equitable and sustainable wellbeing. Raising eco-efficiency within a green growth framework is the first step, but inadequate. Recomposing consumption raises the stakes and interrogates consumer practices in rich countries but continues to take for granted a growth economy. It is an essential further step, but even this will be inadequate. However, it could provide a transition to a world beyond economic growth where consumption demand is reduced in absolute terms in the rich world. I relate these three strategies back to the broader political economies in which they must sit and ask how far they are compatible with different varieties of capitalism or with capitalism as a whole.

6. Decarbonising the economy and its social consequences

This chapter begins with a survey of programmes to reduce territorial emissions in the global North. It focuses on climate mitigation policies. The aim is to get to grips with some detail on decarbonisation programmes – or their absence – in order to assess their social and distributive impacts, and the challenges they pose. It charts some of the distributive and social consequences of these policies and the role of social policies in compensating for them. It then turns to more integrated policies to achieve both social and climate goals, focusing on housing and retrofitting.

Building on the discussion of green growth in Chapter 3, I draw on a range of economic theory, from orthodox neoclassical to heterodox theory. One approach identifies three 'pillars' embracing different conceptions of risk, fields of theory, economic processes and implications for public policy (Grubb et al. 2014). The first and dominant pillar is the economy as conceived by neoclassical economics and optimisation theory. The key policy solution to address climate change is to raise the price of carbon so as to impose costs on polluters and thus internalise the externalities they generate, and provide price incentives for sustainable energy. The second pillar draws on behavioural economics to recognise that many consumers embrace acceptable, achievable goals and will be affected by factors other than price, for example in acting on the energy efficiency of their housing. The dominant policy solutions that follow from this are enforcement of *public standards* and citizen engagement. The third, more radical pillar draws on evolutionary economics, complexity theory and political economy to understand the transformation of whole economies, which are seen to be shaped by the interplay of technology, infrastructure, institutions and interests in historical time. Here, the policy imperatives are to reshape the path of development through *transformative investment* by large private sector organisations alongside strategic public planning and infrastructure investment.

To bring down carbon emissions fast, all three approaches are needed, not just price and market adjustments. There is no silver bullet.[1] These three policy pillars structure the first section of this chapter.

---

[1] There are similar wide disparities in theoretical approaches to public policy-making.

CLIMATE MITIGATION: A FRAMEWORK AND THREE POLICY PILLARS

The scope of potential climate strategies is vast and there is rich literature on it that cannot be encompassed here. To present the major features of it, I discuss, first, the role of legal and institutional frameworks and then examine the three core policy pillars.

Legal and Institutional Frameworks

Even most economists now recognise that the process of radical and fast decarbonisation will require strong 'steering' by governments and states. In the face of regular parliamentary elections, business pressures and other factors favouring short-termism this requires a robust legislative framework and a credible roadmap of targets (Fankhauser 2012). The UK Climate Change Act 2008 is often seen as a model. Enacted by a Labour government but passed almost unanimously in Parliament it set out a target to cut UK emissions by 80 per cent between 1990 and 2050. It also committed future UK governments to a series of statutory, five-year carbon budgets that set binding and fast-reducing ceilings for GHG emissions; the fifth budget, for 2028–32, has already been agreed (see Figure 6.1). It established an independent institution, the Committee on Climate Change (CCC), to monitor progress, advise on the carbon budgets and make recommendations to governments, and provides a mechanism for judicial review should a government systematically disregard these commitments.

This is not a foolproof framework. As I write, the Act is being challenged by the post-2015 UK Conservative government in numerous ways: it has removed Renewables Obligation subsidies for onshore wind farms and reduced them for biomass, extended the Climate Change Levy to renewable power (turning it into an energy levy), scrapped low-carbon Vehicle Excise Duty, provided new tax reliefs for oil and gas exploration, and fast-tracked fracking. Climate denial remains on the political agenda (Maza 2016). But the government has not yet reneged on the Climate Act,

---

Economics has developed an elegant theory, summarised in 'Tinbergen's rule', that the number of policy instruments should not be less than the number of policy targets or goals. The approach is parsimonious, prescriptive and mainly focused on economic efficiency. The political science approach is empirical, recognising that in practice policy instruments reflect contexts, interests and power, are multiple and are in flux (Jordan et al. 2011; del Rio and Howlett 2013). This approach studies policy mixes where policy instruments interact in the form of conflicts or synergies. In studying a messy, 'wicked' problem like climate change mitigation it is apparent that the political science approach is more helpful, and is the one used here.

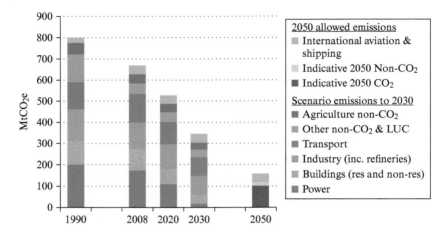

Source: Committee on Climate Change (2010, fig. 8).

Figure 6.1 Indicative UK decarbonisation roadmap

and in 2016 it confirmed the targets recommended by the Committee on Climate Change for the fifth carbon budget for 2028–32. The Act continues to provide a default commitment that places obstacles in the way of a denialist or short-termist government. The framework may also help build a broad political consensus, including some sectors of business, to oppose attempts to derail climate mitigation.

The broad aims of the Climate Act need translating into sectoral targets and plans. One of the earliest documents issued by the new UK Department of Energy and Climate Change in 2009 was an ambitious *Low Carbon Transition Plan* (HM Government 2009), covering power and heavy industry, transport, homes and communities, workplaces and jobs, and farming, land and waste – a remarkable return to the idea of national economic planning. Though the government no longer follows the plan as a policy guide the CCC continues to publish sector-specific reports and roadmaps. Figure 6.1 reveals the scale of the challenge facing the British economy, which serves as an illustration of the challenges facing all developed economies. To hit the 2050 target, carbon emissions must be cut by 3.2 per cent year on year up to 2030, and this still leaves a faster rate of reduction to be achieved in the following two decades up to 2050.

As Figure 6.1 shows, by 2030 the brunt of planned decarbonisation will be borne by the power sector and (to a lesser extent) by buildings, both commercial and residential. Transport, industry and agriculture are expected to make much smaller contributions, and international aviation and shipping are excluded entirely. In its report for 2014 the CCC sees encouragement in the scale-up in deployment of wind generation, both onshore and offshore, but recognises serious shortfalls in most key areas, including roll-out of technologies such as carbon capture and storage, electric vehicles and heat pumps. It notes dramatic slowdowns in energy efficiency measures in residential buildings, transport more generally, industry and farming. The Committee on Climate Change (2014) doubts that present policies are adequate to achieve the broader goals.

Carbon Pricing

Any market economy needs to raise the price of carbon as part of a sustainable climate policy in order to internalise the huge external costs of climate change and to incentivise actors to reduce emissions.[2] Standard economics recognises the need for some public policy, but defines its role as rectifying significant market failures, notably *externalities* – unintended consequences of market actors with direct, usually negative, impacts on the welfare of other consumers or the costs facing other firms. To internalise the costs of emitting GHGs the theory favours a higher, undifferentiated *carbon price*. This will feed its way through the economy, altering the relative prices of the plethora of goods and services to which different firms and consumers will respond. It offers firms and consumers greater flexibility in deciding how to adjust. The result is claimed to be more efficient carbon mitigation.

There are two ways of raising the price of carbon. Carbon taxes directly raise prices and leave markets to decide how much carbon emissions will be reduced. Cap-and-trade schemes limit the amount of carbon to be emitted and distribute emission permits to firms that they can then trade, which indirectly determines the implicit price of carbon.

Little has been achieved in the way of carbon taxation, despite the strong case for shifting the tax burden from goods and labour to 'bads' like carbon and pollution. After a brief surge in the 1980s to early 1990s the level of explicit carbon taxes has waned ever since (Environmental Tax Policy Institute and the Vermont Journal of Environmental Law 2008). Six EU countries have enacted what can be described as environmental tax reform (ETR) since around 1990: Denmark, Finland, Sweden, Germany, the Netherlands and the UK. Modelling the effects shows both environmental and economic outcomes have been broadly positive: energy demand and emissions reduced, employment increased and the effects on

---

[2] Thus I disagree with those who claim that all such price-based instruments are 'neoliberal'.

GDP and industrial competitiveness minimal (Andersen and Ekins 2009). However, the politics and the distributional impacts are problematic, as discussed below.

More usually, specific taxes, such as petrol and diesel duties, act in a similar way, though they were often introduced to meet other goals. Throughout the OECD the implicit carbon tax on transport emissions is far higher – 13 times higher on average – than the taxation of heating and electricity, which remain nearly tax-free. Indeed, in the UK, domestic energy is exempt from standard VAT and charged at a much lower rate of 5 per cent. This variation is another argument for a more systematic carbon tax.

The dominant approach has been cap-and-trade schemes. These set a target level or cap for emissions to which the price of carbon will adjust – as opposed to carbon taxation, which raises the price of carbon and thus influences the quantity of emissions. Though under certain assumptions carbon taxes and cap-and-trade can be shown to deliver identical outcomes, in practice they differ in several ways. Notably, carbon trading creates new property rights in the emission of GHGs – a new market that fosters rent-seeking and gaming by corporations, financial institutions and other market actors.

The EU Emissions Trading Scheme (ETS) is the world's largest cap-and-trade system, applying to power generation and large industrial concerns across Europe. Companies can trade their allowances and also use credits from economies achieved in developing countries. Yet thus far the ETS has failed to achieve even the most modest rise in carbon price: the current ETS price is €8 a tonne, whereas one calculation of the price necessary to achieve UK carbon targets at present is €83 a tonne – ten times higher. Given its ineffectiveness, a plethora of other national schemes have emerged in EU member states; in the UK these include the Carbon Floor Price, Climate Change Levy, Carbon Reduction Commitment and Renewables Obligation.

Taking account of all policies, the OECD (2013) has estimated *average implicit carbon prices* and finds a wide variation across the major countries and regional groups (Table 6.1). The contrast between Europe and the US is startling; Japan lies between the two. The UK is in line with the leading states of Europe, notably the Nordic countries, Italy and Switzerland. As yet there appears to be little convergence across the EU in carbon-pricing policies.

Cap-and-trade has many weaknesses. Once established, schemes such as the ETS create opportunities for gaming and vested interests, which then lobby heavily for its continuation and for less restrictive implementation. Futures trading in carbon also generates a powerful financial industry

Table 6.1   Average implicit carbon prices in major OECD countries

	£/tCO$_2$
UK	59
Germany	47
France	50
Italy	62
Sweden	64
EU	53
US	4
Japan	30
OECD	47

Source: Bassi et al. (2014).

interest. Industrial lobbying and 'lock-in' then raise the costs of changing the implementation of cap-and-trade, let alone moving towards carbon taxation. The outcome, according to the UK National Audit Office (2009), is an economic and regulatory landscape so complex that it is impossible to evaluate. Yet despite its inherent problems it has prevailed over carbon taxation – according to Helm (2008) and Hepburn (2009) because it reflects the hegemony of market mechanisms within current neoliberal thinking.

Public Regulation

The second pillar of climate mitigation policy includes publicly authorised and implemented *standards* and *regulations* to improve energy and emissions efficiency. These include 'product regulation' covering vehicles, appliances, the energy performance of buildings and so on. Examples are the EU ban on incandescent lights, vehicle emission standards, Standard Assessment Procedures (SAPs) for buildings and tighter building regulations. Government standards and regulations have proved more effective than voluntary programmes. The EU labelling of electrical appliances has transformed their average energy efficiency over the past two decades. Fears expressed by business groups have largely proved groundless, and energy efficiency policies once in place have rarely been reversed. The EU imposes roughly common standards across all member states; by providing common treatment of industries it demonstrates the benefits of wider regional regulation and standard-setting. Indeed public regulation and standard-setting is the unsung success story of decarbonisation. Despite this record they are often ignored or downgraded in much economic analysis, primarily because they run counter to neoclassical theories of consumer and business sovereignty and lack potential for profit-making (Hills 2012; Grubb et al. 2014, ch. 5).

However, a recognised side-effect of all such non-monetary policies to change energy and emissions behaviour is 'rebound'. Rebound occurs when, for example, saving money on energy through efficiency standards in one area results in expanding consumption and emissions in another, whether directly (e.g. turning up thermostats) or indirectly (consuming more of other carbon-intensive goods). Recent estimates suggest that the problem is not overwhelming in the UK: rebound in households varies between 10 per cent and 30 per cent, which means that the energy and emissions saved amount to between 70 per cent and 90 per cent of the initial savings (Chitnis et al. 2014). However, it strengthens arguments for a collective policy on consumption: the extent of rebound varies according to the degree of engagement people have with energy efficiency policies. Citizens engaged through local participation can better appreciate the potential for rebound and act to avoid it (Grubb et al. 2014, ch. 5). I return to these issues in the next chapter.

Green Investment

The third decarbonisation pillar involves strategic, transformative investment in low-carbon and zero-carbon technological innovation, infrastructure, energy supply, energy networks, transport and building stock, as well as investment in adapting to climate change. This approach rests on a distinct political economy paradigm. Investment in technical and organisational innovation is critical to bring about a green transformation, and this will require substantial and sustained public action, such as public banks, state support for R&D and university research, and a coherent industrial policy. The state can deliver 'patient capital' to augment and control the hot short-term capital often rules in financialised capitalism. This approach rejects the primacy of markets, regarding them instead as continually co-created by states. It insists that economic value is created by both private and public sectors, often in collective collaboration. Given this collective value creation, it is argued that profits should not be monopolised by the private sector, but that the state and citizens should share in the proceeds (Mazzucato 2011; Jacobs and Mazzucato 2016).

'Green' investment has been growing over the past two decades, but it falls well short of what is needed to attain the 2 °C target. It is estimated that a further \$650–\$900 billion will need to be invested each year up to 2050, equivalent to some 2 per cent of global GDP *per annum* (IEA 2014). This investment backlog is not just found in the developing world;

indeed, the transformation of inherited capital stock and networks is especially difficult in developed economies where high-carbon systems are 'locked-in'. Yet there are enormous problems of institutional alignment, conflicting incentives, politics and short-termism to overcome.

Both private and public investment would be required, but private green investment faces structural obstacles. The aftermath of the 2008 financial crisis continues to depress expectations and favour low-risk investments. High fixed costs render marginal cost pricing unprofitable. The system and network properties of much technology, energy and transport require coordination and planning. Countries with large public green banks, such as Germany, China and Brazil, have overcome some of these problems, but in all countries they are denied the ability that commercial banks have to create credit (Campiglio 2015). In the post-crisis period of high savings, low investment and low interest rates, the case for public strategic investment grows (Zenghelis 2012).

Yet this directly contradicts conventional economic thinking and policy trends. Public borrowing for investment is rarely if ever explicitly shown in national accounts; the present practice of measuring *gross* public debt (as with *G*DP) makes nonsense of the need to provide assets for future generations (Helm 2011). So, while green investment should ideally be supported by public credit banking, it must usually rely on risk-averse private funding. In all advanced democracies policy has shifted towards privatisation, marketisation and deregulation across a wide range of key infrastructure sectors including energy and other natural monopolies. Liberalisation has been extensive in all areas of economic interventions, and especially in energy, and especially in the UK (Höpner et al. 2014). In consequence, net public wealth in the OECD, after being positive from the end of the First World War to 1980, has fallen to around zero in the last three decades (Piketty 2014, ch. 3).

There are parallels between calls to reinstate a coherent public investment strategy to deal with climate change and the 'social investment' approach to social policy experimented with in parts of Europe over the last two decades. Fostering a 'new green industrial revolution' will call for more public investment in research and development, universities and higher and technical education, but there is no doubt that this in turn will require more investment in earlier years of education and other public services designed to build human and social resources.

Combined Effects of Mitigation Policies

The very fact that climate mitigation policies (CMPs) are being developed (to differing extents) in all countries buttresses the argument that carbon

Table 6.2   Per capita territorial emissions in 2030 for INDC scenarios

	Emissions per capita 2010	INDC scenario(s) for 2030	Change 2010–30: absolute changes	Change 2010–30: % changes
US	17.3	10.6–11.3	−6.7/−6.0	−39/−35%
EU	7.2	6.1	−1.1	−15%
Other OECD	8.2	7.9–8.7	−0.3/+0.5	−4/+6%
China	5.3	9.0–11.5	+3.7/+6.2	+70/+117%
World	4.3	6.3–7.3	+2.0/+3.0	+45/+68%

Source: Boyd et al. (2015, table 7).

mitigation is not simply a 'burden' that rational business actors in the North will only accede to as a result of international agreements foisted on them. There are dynamic internal reasons for expecting national mitigation strategies to take off. This explains the existence of national carbon commitments over the past two decades, now expressed in the INDCs put forward at the Paris COP. National carbon plans and budgets do not necessarily undermine competitiveness in the global economy (Bassi et al. 2014).

Another way to summarise current mitigation efforts is shown in Table 6.2, which focuses on planned climate targets. It presents current INDC target ranges for territorial emissions per capita for 2030. If these targets are achieved then the current very wide gap between the North, East and South should be narrowed by 2030. Territorial emissions per capita in the US and China will reach rough parity, the EU will continue to reduce emissions but at a slower pace, while the rest of the OECD will fail to cut significantly. But global emissions per capita will rise by between one-half and two-thirds in the next decade and a half. The planetary gap between national plans and the needed target of 2 tonnes per person will continue to widen.

SOCIAL AND DISTRIBUTIVE IMPACTS OF CARBON MITIGATION: DOMESTIC ENERGY AND HOUSING

To what extent might the pursuit of carbon mitigation undermine the pursuit of social or distributive goals? *Social* impacts refer to any effect on components of human need satisfaction and wellbeing, while *distributive* impacts refer to differential effects across different social groups. Are

CMPs necessarily regressive in their effects? Can some decarbonisation strategies secure both environmental and social co-benefits? The diversity of mitigation policies, their situation within broader economic, social and climatic regimes, and the interactions between all these things makes modelling these impacts a complex task. To render it manageable and to identify clear links between climate and social policy, I consider here only the impact on housing and household energy consumption.

Housing, for long a marginalised area of social policy, is now coming back centre-stage, owing in part to its intersection between social policy and climate policy. Around 25 per cent of total emissions in OECD countries come from housing and the energy used to heat, cool and light dwellings and run the growing range of electrical appliances. The potential contribution of domestic energy efficiency to climate mitigation is regularly trumpeted, especially since many of the policies are in theory 'win–win–win' – reducing emissions at the same time as cutting household energy bills and improving comfort levels in the home. Yet many countries have failed to restrain housing emissions (Eurostat 2016).

Housing has since the nineteenth century been a target of social policy, variously for its adequacy, quality and cost, and for its impact on health, environment, poverty and social exclusion. The forms of public intervention have varied greatly across OECD countries. Clear differences exist, for example, in the extent to which homes are owner-occupied or rented, and whether housing finance for owner-occupation or the rented sector is regulated or liberalised. Like all policy fields, housing has been deregulated over the past three decades, but this has not yet resulted in convergence between countries. A range of 'housing regimes' have been identified across the OECD.[3]

It is reasonable to suppose that these system differences will affect the energy use and emission patterns of households across countries (Schaffrin et al. 2015). Housing impacts on emissions in at least three ways: via capital investment in new building, via the current energy costs of running the housing stock, and via its indirect effects on spatial relations and such factors as the demand for transport. Here I focus on current energy emissions in the UK, which has the most deregulated energy market in the OECD (Höpner et al. 2014).

---

[3] One model for example distinguishes: liberal market (high owner-occupation and deregulated mortgage markets in the Anglosphere), familial (high ownership but regulated mortgage debt in Italy, southern and eastern Europe), corporatist-market (lower ownership plus liberal mortgages in Germany, the Netherlands and Denmark) and statist (low ownership and regulated mortgage debt in France and Austria). There are also different patterns in regulating the socially and privately owned parts of the rented sector (Kemeny 1981; Schwartz and Seabrooke 2008).

As in most rich, growing economies, UK housing energy use increased between 1970 and 2010 for a variety of factors: a relatively fast-growing population, a 40 per cent rise in the number of households, smaller households using more energy per head, a still faster rise in 'under-occupied households',[4] the spread of central heating (an average temperature rise of 3.6 degrees over the last four decades in the UK), and many more electrical appliances. Despite this, energy bills fell up to 2004 and overall emissions declined a little, owing mainly to the switch to gas and phasing out of coal and oil (DECC 2010). Yet the standard of home energy efficiency remains low in the UK, and the Committee on Climate Change (2016) warns that current plans will not achieve the carbon targets in Figure 6.1.

I consider first the distributive implications of *carbon pricing* and the countervailing redistributive measures that have been suggested and put in place.[5] The discussion is then broadened out to consider the impacts of *green new deal* investment policies on a wider range of social impacts including employment. This introduces the idea of 'co-benefits' and of more unified 'eco-social' policies.

Pricing Domestic Energy and Energy Poverty

Raising the cost of carbon – through taxation and/or cap-and-trade schemes – is an essential component of decarbonisation policy, yet it always has a regressive impact. This follows from the fact that housing and energy are basic human need satisfiers. A certain minimum consumption of both housing and domestic energy is necessary to ensure an acceptable level of health, wellbeing and social participation in all societies, though the detailed requirement – the housing need satisfiers – will differ across societies. Because of this need-based element, expenditure on domestic energy varies less across income groups than for any other major consumption category, which means that it accounts for a rising share of income as household income declines.

Figure 6.2 shows that spending on domestic energy rises with income, but that the lowest-income group spend the highest proportion of their disposable income on it: 8 per cent compared with 3 per cent in the top-

---

[4]   Under-occupied households in the UK are usually defined as having two or more spare bedrooms. The numbers rose 45 per cent between just 2003 and 2008 (DECC 2010). The 'environmental impacts of divorce' have been analysed by Yu and Liu (2007). Another distributive issue is the growth of second homes.

[5]   Raising standards through the second pillar – direct regulation – also incurs costs and has distributive consequences (Advani et al. 2013). However, these are less well researched and will tend to be shorter-term once the cost of higher-efficiency appliances comes down through competition.

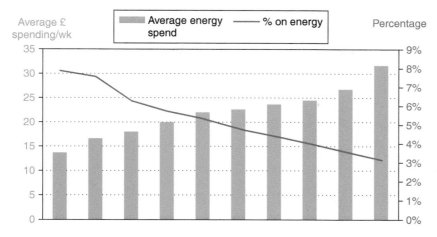

Source: DECC (2013).

Figure 6.2   Average UK weekly expenditure on fuel, light and power, and income, 2011

income group. Small, notably single-person, households and 'workless' households (retired, unemployed or 'unoccupied') also spend a higher share on energy. Thus any charge on carbon will impact more heavily on lower-income households via higher energy bills. There is evidence too that poorer households on average spend more time at home and thus have greater energy needs (Druckman and Jackson 2008; Gough et al. 2011).

The combined effects of social need, unequal incomes and the market pricing of energy can lead to 'energy poverty' or 'fuel poverty'. Since 2014, fuel poverty has been defined in England (not the rest of the UK) as a situation 'where a household has required fuel costs above the median and where to spend that amount they would be left with a residual income below the official poverty line' (Hills 2012). 'Required fuel costs' – the need element – are those needed to heat the home to an adequate standard (currently deemed to be 21 °C in living room and 18 °C in other occupied rooms in daytime hours). Fuel poverty is determined by three factors: energy costs, household income and the energy efficiency of dwellings.

According to this measure 10.4 per cent or 2.3 million households in the UK were in fuel poverty in 2013.[6] They may be faced with a choice

---

[6]   In Northern Ireland a remarkable 42 per cent are in fuel poverty, though on a different basis of calculation.

between reducing heating and reducing other spending levels, either of which can undermine health and wellbeing. A prominent feature in the UK is 'excess winter deaths', some 27000 each year over the last decade, reflecting a wider pattern of ill-health due to cold and damp. This is a far larger peak than in many other countries with colder climates and is generally attributed to the poor energy efficiency of the housing stock. Some households, particularly elderly, disabled and more housebound groups, can suffer from a lack of 'thermal comfort' at odds with societal expectations. Meanwhile fuel poverty means less money to spend on other basic necessities such as decent food: a choice between 'heating and eating'.

Similar concerns can be found in some other countries, such as 'energy precarity' in France, defined as 'encountering particular difficulties in accessing the necessary energy supply to meet basic needs' (Day et al. 2016). In Germany, 17 per cent of households were estimated to be in energy poverty in 2008, defined as having to spend more than 10 per cent of income on 'required fuel costs' (Kopatz et al. 2010). Partly this reflects rising electricity prices associated with the federal energy transition policy (*Energiewende*), a contradiction discussed further below.

But these are averages, and another feature of domestic energy is the great variety of household emissions within each income group, owing to a host of other factors affecting energy use, including the energy efficiency of dwellings (Roberts 2008). This makes it difficult to track the effect of any increase in energy costs; the energy-poor will not necessarily be income-poor, and income-poor households may not be energy-poor. Furthermore, carbon pricing may not achieve a reduction in energy use among higher-income households if the price rise is insufficient to modify their behaviour. Raising the cost of carbon may well lower emissions necessary to provide basic heating whilst having little or no effect on 'luxury emissions' caused by heating or cooling large properties, floodlit gardens and so on.

Since unregulated market economies generate highly unequal income distributions, there is a contradiction at the heart of the orthodox economics paradigm. Put simply, the price mechanism works less well as an allocative mechanism the greater the degree of inequality. The economist Martin Weitzman (1974) writes: 'The price system has greater comparative effectiveness in sorting out the deficit commodity and supplying it to those who need it most when wants are more widely dispersed or when the society is relatively egalitarian in its income distribution.' Where this is not the case, rationing is more effective.

Tackling Inequity: Countervailing Social Policies

Assuming that more carbon taxation and higher carbon pricing are a necessary part of the mitigation toolbox, how – if at all – can these regressive distributive effects be dealt with? There are three basic alternatives: compensate poorer households, vary energy prices, or develop energy efficiency programmes and target them initially at poorer or fuel-poor households (Hills 2012).

Compensation
The standard economics response to these distributive impacts is to 'compensate the losers'. Compensation policies can take a variety of forms depending on the welfare and housing regimes of different countries. In the UK the only explicit mechanisms are *Winter Fuel Payments* and *Cold Weather Payments*.[7] These are remarkably poorly targeted, benefiting only 10 per cent of the fuel-poor while disbursing money to well-off retired households (Boardman 2010, ch. 3).

Can more equitable forms of compensation be devised? All studies so far suggest that, while losses due to higher carbon prices or taxes can be compensated for across income groups, large numbers of households continue to lose out. This is because the variables affecting domestic energy efficiency cannot easily be addressed by existing social transfer programmes, since they encompass factors such as the energy efficiency of dwellings, urban–rural differences, and connection to the gas network.

These problems would multiply in the UK if the present low costs of domestic energy were to be seriously tackled, for example by raising the 5 per cent rate of VAT on domestic fuel to the standard rate (currently 20 per cent) as in most EU countries. A study that modelled the effect of raising VAT and applying the Carbon Price Floor to domestic gas found that this could be compensated for by increasing the basic amounts of Universal Credit and reducing its withdrawal rate, without too many low-income households losing out. Yet the social security costs would be expensive, and they would rise year by year as more ambitious carbon reduction targets kicked in. More people would be subject to means-testing, so facing high marginal tax rates if their incomes increased; yet some low-income losers would remain. Most problematic, the compensation costs would soon exhaust the extra revenue raised by environmental taxes, leaving nothing to spare for improvements in energy efficiency.

---

[7] A flat-rate payment of £250 to households with pensioners (£400 if over 80), and additional payments to pensioners and low-income households during exceptionally cold weather.

The consensus among social policy experts is that compensation plans are both technically and politically very difficult to achieve. Dealing with averages has political implications. Substantial numbers of low-income losers from these policies can spark a backlash against all climate policies. It is not enough to compensate households on average; those most adversely affected need to be addressed in particular. But that is almost impossible to do using compensation policies when dwelling spaces are so heterogeneous.

The same arguments apply to more radical tax plus compensation packages, such as James Hansen's 2009 'carbon tax and 100 per cent dividend' proposal for the US. This would derive the revenue from a carbon tax, while other similar schemes would raise revenue from auctions under a cap-and-trade scheme, such as Cap and Share and Cap and Dividend (Büchs et al. 2011). But all envisage pricing carbon to raise substantial revenues that are then returned in whole or in part to citizens, usually via a lump sum. Given the skewed distribution of carbon footprints, more than 50 per cent of the population would benefit from such a scheme. In this way, it is claimed, carbon mitigation and progressive income redistribution would both be served. The schemes resemble proposals for a basic or citizens' income, discussed in Chapter 8. But alongside well-known critiques of basic income, the evidence presented below clearly demonstrates that *investment* in energy saving and retrofitting achieves much greater carbon benefits in the medium term without necessarily damaging low-income households. Decarbonisation requires a massive rise in strategic investment, not the disbursement of revenues to support consumer spending on energy. Compensation cannot work, so alternatives must be found.

Social tariffs

An alternative way of ameliorating the regressive effects would be *social tariffs* that adjust energy tariffs in line with energy need. This can be done on a targeted basis (for example the UK *Warm Home Discount*), but it is costly and again reaches only one-quarter of the fuel-poor. To extend it to all households on means-tested benefits would entail using complex data-matching and would be 'extremely challenging' (Hills 2012).

A simpler and more universal scheme would require energy companies to operate a 'rising block tariff', with lower tariffs for initial units of electricity or gas consumed, and higher tariffs for successive units. At present energy tariffs work in the opposite way, mirroring the cost structure of utilities faced with large fixed costs to maintain the network. To reverse this would recognise the basic need component of the first block of household energy and the choice element in successive units. The total average price of domestic energy would continue to rise over time, as part of the carbon pricing strategy, but the distribution of the burden would be skewed more to higher-consumption households. Allowance would need to be made for different-size households in line with findings of research on the comparative burdens of energy bills. In principle this could directly link environmental and social goals. Related tariffs schemes have been implemented in some countries in southern Europe (Schaffrin 2014). It would however be difficult to administer in the privatised energy system of most countries today and is opposed by energy suppliers.

In effect this approach would extend the range of goods that are subject to some measure of non-price allocation: it would extend the welfare state to 'socio-natural resources' like energy and water (Fitzpatrick 2014a). Policies to bring the allocation and pricing of energy back to some form of collective control would challenge the privatised ownership of energy supply.

Targeted energy efficiency programmes
The logical alternative is to act on the third driver of fuel poverty (other than energy prices and household incomes): the energy efficiency of dwellings. To tackle inequality and energy poverty these will need to be either very large-scale, approaching universal coverage, or targeted on households with greater need, such as low-income families and/or energy-inefficient housing. There is a wide range of such programmes across countries but only two basic ways of funding them – government taxation or raising average tariffs on energy consumers.

Until 2013, the UK provided government subsidies for thermal efficiency, notably the *Warm Front* and the *Decent Homes* programmes, targeted on social housing and low-income households. These were abolished in 2013, so that reliance is now placed on mandated energy company programmes. For example, energy companies are required, under the new Energy Company Obligation (ECO), to provide boiler upgrades and insulation to low-income housing, deprived areas, and hard-to-heat housing of the poorest and most vulnerable households. The result is a suite of mandated private programmes to aid poorer and less well-housed groups. Yet because these are paid for out of higher energy bills they again paradoxically *increase* rather than reduce fuel poverty (Hills 2012).

In his independent review of fuel poverty, Hills (2012) compared these three approaches and came to some clear conclusions. Energy efficiency measures are superior to other policies on all counts – environmental, social and distributive. Compensating low-income households via more income is the least successful or efficient (Gough 2013).

Green Investment: Retrofitting Housing for Energy Efficiency

Our attention must therefore turn to the third pillar of carbon mitigation: large-scale, publicly led investment. Much of this will take place upstream, notably investment in renewable electricity generation, but again our attention here is on downstream investment, in this case in retrofitting the housing stock. The Committee on Climate Change (2015) in the UK now calls for additional tax-funded measures to rapidly improve domestic energy efficiency: retrofitting the entire housing stock is the urgent priority.

The most embracing proposals since the 2008 financial crash have been various forms of 'green new deal' (NEF 2008; Barbier 2010). These call for a sustained public programme to invest in renewable energy and to deploy radical conservation measures. This would at the same time boost demand in slow-growing post-crisis economies and create new employment opportunities, via, for example, creating and training a 'carbon army' of workers to achieve the reconstruction and house retrofitting programme. It is a radical exercise in policy integration for a post-crisis economy.

The UK provides supporting evidence when modelling the impact of a suite of carbon taxes offset by reductions in employers' national insurance contributions (which are effectively a tax on labour). The reduction in labour taxation would create an estimated extra 455 000 jobs by 2020, and even more if part of the revenues were to be invested in making homes more energy-efficient.

At the EU level it is claimed that an investment programme of 2 per cent of EU GDP per annum would create 3 million jobs, split roughly three ways between jobs in renewable energy, energy efficiency measures and transmission infrastructure. Finance would come from tapping pension funds, insurance funds, mutual funds and sovereign wealth funds, rather than from commercial banks (World Economic Forum 2013). The general case for a green growth strategy has been made and discussed in Chapter 3, and this is one concrete example of that. It has made little headway in national policies, with a few notable exceptions.

For example, Scandinavian countries have implemented strong building standards and have built widespread combined heat and power systems. Germany has the most ambitious energy-saving programme in Europe, aiming for a 30 per cent reduction in energy usage by 2020, and a 30 per cent renewable energy share, consisting mainly of biomass, wind and solar (Power and Zulauf 2011). Germany's energy-saving programme is based on three pillars. First is a clear legal framework and tight regulation at the national level, requiring energy efficiency upgrades to buildings and increased use of renewable energy sources among electricity providers. Second, at all levels of government strong financial incentives are provided through subsidies and loans to reduce energy consumption in the built environment; at the national level, these are provided via a public investment bank sponsored by the German government. Third, there is coordinated information, promotion, and behaviour change, working through regional and local bodies, developing enforceable standards through energy performance certificates, accrediting installers and supporting model projects all over Germany. Hence, the strategy covers all three policy pillars outlined earlier.

Since 2006, Germany has created nearly 900 000 jobs in retrofitting homes and public buildings such as schools. To date 9 million homes have been retrofitted to high-energy efficiency standards, and the annual rate of refurbishment is around 400 000 units each year. Having taken these steps, Germany remains on track to meet aggressive greenhouse gas reduction targets by 2020 and 2050. The cost of interest-rate and investment subsidies is considerable, but it has been estimated that every €1 of subsidy leverages in €9 in loans and private investment (Schröder et al. 2011). Owing to the feed-in tariff (FiT) – paying households that install photovoltaic panels on their property a favourable tariff for the electricity they supply to the grid – there are now 700 000 energy suppliers in Germany. Thus a new political constituency is emerging that can challenge the central energy companies and fossil fuel interests.

These are clear examples of *eco-social* policies, policy instruments that serve two goals: environmental in reducing emissions, and social in recognising and meeting a basic need, reducing energy bills and building useful employment. Even so, there are short-term regressive impacts: the FiT programme clearly benefits households and farmers in the top 30 per cent of income distribution (Hüther 2013). Since FiT is financed in large part through higher electricity prices the overall impact is regressive. However, in a dynamic context, the FiT has proved invaluable in building up Germany's solar panel industry.

CONCLUSION: EQUITABLE ECO-EFFICIENCY

To decarbonise national economies, the three pillars of carbon mitigation described at the outset of this chapter need to be combined and aggressively pursued. Raising the carbon price is one essential element, but only one. In the field of housing and domestic energy direct regulation of standards and a strategic green investment strategy are even more important.

To ensure that the overall outcome is not regressive and does not harm other aspects of human welfare, compensation policies will not be sufficient and could be counterproductive. New forms of 'eco-social policy' are required, notably a comprehensive green new deal coupled with social

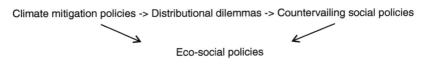

Climate mitigation policies -> Distributional dilemmas -> Countervailing social policies

Eco-social policies

Figure 6.3   From social policy to eco-social policy

energy tariffs. The model of social policy needs to change, as illustrated in Figure 6.3. Instead of reactive and countervailing social policies, there is a need for proactive, integrated 'eco-social' policies.

These, together with the new roles required of existing social policies, are summarised in Table 6.3. This builds on Table 5.1 in the last chapter in order to show how the social/equity dimension can be combined with improving eco-efficiency.

However, carbon mitigation policies, like social policies, are not lists of options to be drawn on or rejected by policy-makers. They are situated in socio-economic-technical-ecological systems inhabited by specific sets of actors and displaying considerable path-dependency. For example, the UK, despite having bold climate targets, remains a laggard in reducing emissions from domestic energy for an integrated set of reasons. A recent summary of the 'heating regime' in the UK concludes:

Table 6.3   Summary of eco-social policies in Chapter 6

	Welfare goals	CI: Ramp up eco-efficiency
S1: Redistribution	Guarantee a minimum-income floor. Reduce inequality.	Where possible, compensate carbon-pricing losers. Reduce inequality to enhance the effectiveness of carbon pricing.
S2: Social consumption	Collectively provide vital need satisfiers. Regulate and prevent harmful consumption.	
S3: Social investment	Develop human and social capabilities.	Build personal and social capabilities for green new deal programmes.
ES: Novel eco-social policies		ES1: Green new deal: house retrofitting and allied programmes. ES2: Social tariffs for energy and water.

The heating regime is fairly stable in particular due to strong infrastructural lock-in (gas grid/inherited housing stock), the concentration of powerful actors on the supply side, the captivity and relative lack of awareness on the demand side (consumers), and a tendency for business as usual in the equipment installation and maintenance trade. (Turnheim et al. 2015)

This regime is linked to the political economy of the UK and the power of incumbent energy actors within it (Geels 2014; Ellis and Barry 2015).

Radical carbon mitigation will not be achieved through market mechanisms. Nor could it be achieved at any sufficient scale or speed by locally determined initiatives. It requires a shift from the neoliberal model towards a more dirigiste, state-led strategy: an active interventionist 'innovation state', with substantial public investment, state banking, subsidies and other incentives to private investment and greater regulation and planning. This would involve at least the following: a national investment framework to embrace renewables and innovation in electricity, gas, transport, urban and electronic infrastructure; further social investment in education, labour market access, communities, and health including preventive strategies; where there are important natural resources, such as oil, an investment fund to cover the amortisation of reserves; investment in adaptation to climate change; and widespread long-term investment in various forms of conservation and eco-maintenance. The investment functions of social policy would also need to be enlarged and more closely integrated with environmental investment (Jackson 2009, ch. 12; Jacobs and Mazzucato 2016).

Moreover, such combined intervention would require changes in the organisation of government and public administration. Policy would need to be more integrated, both horizontally and vertically. Horizontal integration would connect previously disparate sectors, ministries and agencies, such as energy, social protection and finance. Vertical integration would devolve power and tap into the local knowledge of communities and local government while coordinating these at the national and supra-national level (Lafferty and Hovden 2003). But such administrative reforms would also challenge powerful interests, institutions and ideological justifications. We return to these issues of political economy and path-dependence in the final chapter.

I have concentrated so far on territorial emissions and the climate goal of improving eco-efficiency. Although this is of first-rate importance, it will not be sufficient, because it targets only one variable: decarbonising production. It is time to turn to the sustainability of consumption.

7. Decarbonising consumption: needs, necessities and eco-social policies

> We are persuaded to spend money we don't have on things we don't need to create impressions that won't last on people we don't care about.
> (Tim Jackson)

This chapter turns to the second strategy for reducing emissions and global warming – reducing the carbon and GHG content of consumption. It has been much less discussed in the literature than improving eco-efficiency. It will give us a new perspective on the wellbeing and distributive implications of tackling climate change, and the role of social policy.

I made the case in Chapter 3 for calculating a nation's emissions on a consumption basis rather than a territorial basis for four reasons: ethical, distributional, political and, finally, realist – it is *extremely* unlikely that the eightfold decarbonisation required across the globe can be achieved using just measures to decarbonise production. In the tug-of-war at the global level between rising affluence and population on the one side and improving eco-efficiency of production on the other, the former always wins out except for short periods of crisis and recession.

The same is true within the global North. To illustrate this consider the change in aggregate consumption emissions in Sweden between 1993 and 2008 shown in Table 7.1. Swedish consumption emissions have been boosted by population growth and notably by growth in living standards and per capita consumption; they have been deflated by rising eco-efficiency of production and, to a small extent, by a shift in the composition of consumption. Yet Swedish emissions continued to rise by 0.5 per cent each year.

These emissions will need to be cut drastically by 2050 to achieve global warming targets. One model of the likely alternatives is shown in Figure 7.1. If the rate of improvement of eco-efficiency matched that of the past (1.4 per cent per annum), Sweden's emissions would almost level out (the second line) – a considerable achievement given that the 'low-hanging fruit' has already been picked. But even a doubling of this rate (the third line) would not be enough to achieve the necessary CO_2e target of 2 tonnes per head by 2050.

Table 7.1   *Decomposition of trends in consumption-based emissions in Sweden, 1993–2008*

	Impact on annual change in emissions (holding other factors constant)
Population	+0.5%
Consumption per capita (C3)	+1.8%
Change in consumption composition (C2)	−0.4%
Eco-efficiency of production (C1)	−1.4%
Total emissions	+0.5%

Source: Nässén (2015, table 1).

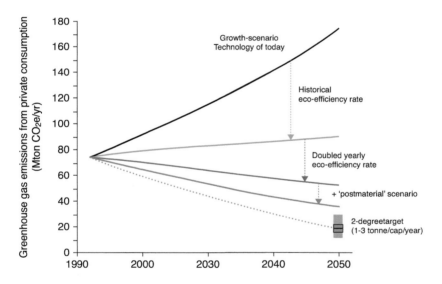

Source: Nässén (2015, fig. 5).

Figure 7.1   *Scenarios for greenhouse gas emission from private consumption in Sweden, 2012–50*

Thus it is highly likely that both 'post-material' and 'sufficiency' policies will also be needed. Nässén's model looks at each in turn. First, a 'post-material' package of policies to change the composition of consumption is modelled including, for example, policies to cut beef consumption, air travel and expensive second homes. He estimates this could conceivably achieve the fourth line of reduction, but they would still not be enough. The 2050 goal is finally achieved by introducing a 'sufficiency' policy to reduce *overall* consumption in Sweden, through a reduction in average hours of work.

These findings show that, in any realistic scenario, simply raising eco-efficiency, though crucial, will not be enough. Consumption itself must be *'recomposed'* to cut emissions – that is, we must change the patterns of consumption (C2). And, if that in turn is not enough, the total quantum of consumption in the North must be reduced (C3). The second option is analysed in the next chapter. Here, I analyse the first.

In this exercise, another goal must have equal weight: equity and the maintenance of some adequate level of wellbeing for all citizens. Recomposing consumption in an unfair way by hitting the poorest and weakest would be unethical, illegitimate and probably counterproductive. As we shall see, the goals of sustainability and equity are not easily reconciled in unequal, highly commodified societies like our own.

Thus we need a conceptual framework that can embrace both wellbeing and sustainable emissions. I have argued that a theory of universal human needs can provide this framework. Since human needs are universal over time and space, and since there are strong moral claims for meeting the needs of all people, they provide a way of understanding wellbeing in rich countries that does not trespass on the need satisfactions of other peoples now or of future generations, as the Brundtland Report advocates.

Need theory enables us to distinguish *necessities* from *luxuries*.[1] To recompose equitably entails making a distinction between goods and services that are necessary for a basic level of wellbeing, and those that are surplus to this requirement. By prioritising the former, need theory provides a bridge to relate social, global and intergenerational justice (Gough 2015a). In the language of Giulio and Fuchs (2014), it enables us to define a 'consumption corridor' between *minimum* standards, allowing every individual to live a good life, and *maximum* standards, ensuring a limit on every individual's use of natural and social resources in order to guarantee a good life for others in the present and in the future.

In this chapter I begin by analysing the size, composition and distribution of all consumption-based emissions in the UK and some other OECD countries. When these are cross-referenced against a rough measure of their necessitousness, an important dilemma is revealed: basic necessities are on average higher carbon than non-necessities or 'luxuries'. I then return to the theory of human need to set out a methodology for identify-

---

[1]   'Luxuries' suggests mansions and private yachts, but I cannot think of a better term to refer to everything that is not a 'necessity'.

ing a minimum bundle of necessary consumption items in the UK and suggest how it might be used to identify a *maximum* bundle for sustainable consumption. I survey corporate barriers and structural obstacles in the path of sustainable consumption, and then propose a further range of eco-social policies to support the recomposition of consumption: taxing luxuries, rationing carbon and socialising consumption. These are combined into an overall strategy for recomposing consumption, based on precaution and prevention. I conclude by summarising how this whole approach challenges some fundamental principles of orthodox economics.

THE SIZE AND DISTRIBUTION OF CONSUMPTION-BASED EMISSIONS

To begin with we need to measure the size, pattern and distribution of consumption-based GHG emissions. The methodology for calculating their size is discussed in Chapter 3 and its Appendix. A sizeable number of studies of the carbon or GHG footprints of households are now available for OECD countries.[2] They find that consumption emissions exceed production in almost every country in the North (see Figure 3.3).

At the household level, direct emissions – mainly household domestic energy use and petrol for private cars – account for a minority, between one-fifth and one-quarter of total private emissions. To concentrate on the direct emissions of households, as we did in the previous chapter, is to give a threadbare and distorted picture of the carbon and environmental footprint of consumption activities in rich countries. The majority of emissions are 'embodied' in other goods and services. Moreover these 'embodied' emissions are rising faster over time as global trading expands (Druckman and Jackson 2009).

Figure 7.2 breaks down all UK emissions into six broad categories. Within private consumption, domestic energy and all travel each account for around one-quarter of GHG emissions. Food, consumables and private services each emit roughly one-eighth. Figure 7.2 also shows emissions from *government* services of all kinds, including health services, education, social services, the military and justice, accounting for 1.8 tonnes per person, or some 12 per cent of the total. I shall concentrate on

---

[2] For example, for the UK: Druckman and Jackson (2009, 2010), Baiocchi et al. (2010), Gough et al. (2011); for the US: Weber and Matthews (2008). Kerkhof et al. (2009) provide a comparative study of four countries – the Netherlands, the UK, Sweden and Norway. Studies differ according to whether they measure CO_2 or all GHG emissions and in many other ways. See Druckman (2016) for a survey.

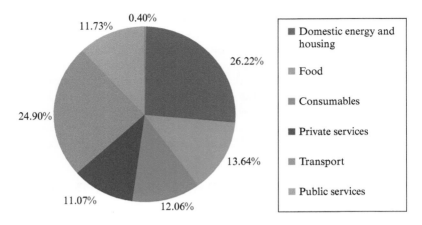

Source: Gough et al. (2011, table 3).

Figure 7.2   Per capita consumption-based emissions by category, UK, 2006

private consumption from now on and return to the issue of welfare state emissions later on.

To discover the *distribution* of consumption emissions across households, this dataset must be linked to another on household expenditure.[3] An example is shown in Figure 7.3. It distinguishes deciles of income, calculated on an 'equivalised' basis to take into account household size and composition. It shows emissions rising in line with incomes; in particular, the emissions of the top 10 per cent are out of line, emitting 5.7 tonnes per person more than the next highest decile. Income is significantly correlated with all types of emissions but to different degrees. Comparing the highest and lowest deciles, we find that emissions per capita for the highest-income group are 4.5 times higher for transport, over 3.5 times higher for private services and consumables and only 1.8 times higher for the more basic goods of domestic energy and food.

As well as income, emissions per head are affected by other factors such as household size, household type, housing tenure, and the employment status and hours of work of the head of the household. Using regression analysis to disentangle these effects we find that income remains the major explanatory factor. Household size varies inversely with per capita

---

[3] In the UK this is the Expenditure and Food Survey. By linking together the data we are able to calculate the average emissions per household and per person for each category of consumption (known as Classification of Individual Consumption by Purpose – COICOP) (Gough et al. 2011).

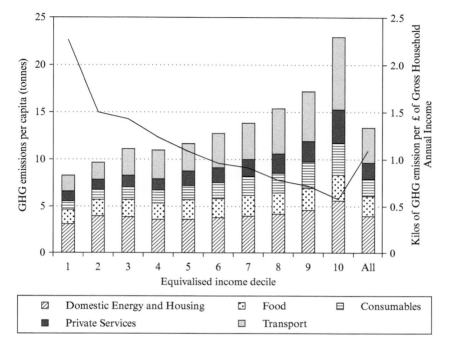

Source: Gough et al. (2011, figs 2, 8).

Figure 7.3 *Per capita GHG emissions, and emissions per £1 of income, by income decile, UK, 2006*

emissions, illustrating economies of scale in consumption: single-person households have the highest per capita emissions, followed by two-person households, followed by households with children. Non-retired 'work-less' households emit significantly lower amounts than working households when income and composition are taken into account (Gough et al. 2011).

Since we are concerned with the distributional implications of policies to reduce carbon emissions, it is also useful to calculate the ratio of emissions to income. Figure 7.3 also presents average emissions per £1 of income for each of the deciles. Immediately the picture of rising lines is reversed. Per capita emissions are greatest in relation to income in the lowest-income decile and fall as income rises: the lowest decile emits four times as much in relation to its income as the highest. The *emission intensity* of all major categories of consumption is less than one; thus any rise

Table 7.2   Categories of personal consumption by necessity and emission content, UK, 2009

	High emission (>1 tonne CO_2e/£000)	Low emission (<1 tonne CO_2e/£000)
'Necessities' (expenditure elasticity <1)	All domestic energy (electricity, gas, other fuels) (26.9%) Food (12.9%)	Communication (1.2%) Alcohol/tobacco (0.7%)
'Non-necessities' (expenditure elasticity >1)	Vehicle fuels (5.0%) Other transport (10.0%)	Clothing and footwear (2.6%) Other housing (2.3%) Furnishings (5.0%) Recreation and culture (8.7%) Restaurants and hotels (5.0%) Private health (0.5%) Private education (0.3%) Miscellaneous (4.2%)

Note: The figures in brackets show the shares of total GHG emissions accounted for by each item.

Source: Chitnis et al. (2014, tables 5, A.5).

in carbon prices, when generalised throughout the economy, will impact on lower-income households more. However, the degree of regressivity varies according to the type of consumption. Expenditures on, and emissions from, domestic energy and food take a proportionately higher share of incomes lower down the income scale than spending on, and emissions from, transport, consumer goods and personal services. As the previous chapter showed, policies that are financed from home energy bills will be especially regressive.

Table 7.2 presents this information another way by tabulating the *expenditure elasticity* of different goods and services against their GHG emissions intensity. Expenditure elasticity, which measures the ratio of the percentage change in expenditure to the percentage change in household income, has long been used as a behavioural indicator to distinguish 'necessities' from non-necessities or 'luxuries'. The former have an expenditure elasticity of less than one, the latter greater than one (Baxter and Moosa 1996). This is shown on the vertical axis of Table 7.2, plotted against the carbon intensity of different goods and services on the horizontal axis, which simply distinguishes items emitting less and more than 1 tonne a year.

This measure of necessity shows as expected that domestic energy and food have low elasticity and thus count as necessities; perhaps less expected, it shows that alcohol and tobacco also have low elasticity. Spending on the remaining goods and services rises faster than income as income increases: transport, clothing, furnishings, recreation and culture, restaurants and hotels, private health and education, indicating non-necessities or luxuries. All transport expenditure is lumped together and appears as a 'non-necessity' in this table, but a more detailed analysis would move some low-elasticity items, such as driving cars for commuting and shopping, into the top-right-hand box.

Table 7.2 reveals an important paradox: necessities are higher carbon than 'luxuries'. The critical basic goods of home energy and food, accounting for two-fifths of all expenditure on average, are carbon- and GHG-intensive, while almost all non-necessities, measured in this way, are low-carbon. There are *no* low-emission necessities apart from 'alcohol' and communication, though communication technology is destined to expand its scope in the future. The table suggests a potential tension between securing emission reductions and ensuring an equitable distribution. It could imply that simply redistributing income to low-income households might, *ceteris paribus*, raise rather than lower total emissions.

However, *ceteris* is rarely *paribus*. Much would depend on where the redistributed income and associated emissions came from. For example, if there were a cap on excessive emissions by the affluent, this could more than offset the higher emissions of low-income households on, for example, heating and food. Lessening inequality can reduce pressures to consume throughout the population as discussed in Chapters 3 and 8.[4] Second, comparative research finds that carbon intensities for housing and domestic energy are much lower in Norway and Sweden, reflecting the extensive presence of district heating, biomass, hydro-electricity and better insulation in the two Nordic countries (Kerkhof et al. 2009). Different forms of technology and infrastructure can thus profoundly improve the eco-efficiency of specific sectors and thus the potential trade-off between equity and sustainability pictured in Table 7.2. The production and consumption domains must be considered together. So far, however, the evidence is that for many countries emissions efficiency in all three basic areas – domestic energy, transport and food – is trailing badly behind the 2050 pathway for 2 °C.

---

[4] Moreover emission elasticities vary and generally diminish with income: the redistribution would be affected by marginal not average elasticities.

NEEDS, NECESSITIES AND EMISSIONS: CHARTING THE CONSUMPTION CORRIDOR

The analysis above takes for granted the behaviour of consumers, the markets for consumer goods and services, and the distribution of household incomes. But markets and consumer preferences are not sovereign. In this section I develop an alternative way of identifying necessities and surplus goods, utilising the methodology for identifying need satisfiers outlined in Chapter 2. There I argued that identifying satisfiers must draw on two forms of knowledge: the codified knowledge of experts and the experientially grounded knowledge of ordinary people. The process to combine the two we call the 'dual strategy'.

What is urgently needed is to identify consumption bundles that a) meet needs for decent living and b) are sustainable over time, in other words to define the upper and lower bounds of the consumption corridor (Giulio and Fuchs 2014). I look at each in turn.

A long history of poverty studies has constructed and measured the components of a minimum acceptable standard of living. Most have recognised this goes beyond just food, clothes and shelter. Since Peter Townsend's work on poverty (1979), there has been wide acceptance that inability to *participate* in accepted social activities is the defining feature of poverty or social exclusion, though what participation entails will be conceived, specified and measured differently in different societies. This analysis has fostered a range of poverty studies to identify a basket of basic goods essential for effective participation.[5] The first to use a version of the dual strategy methodology – 'consensual discussions' among citizens informed at successive stages by expert input – was Bradshaw et al. (2008). The dual strategy methodology is now being overtly applied to estimate 'decent living' standards across all EU member states (Storms et al. 2013). There are now accepted and proven methods of identifying necessities and distinguishing them from luxuries.

The 2014 Minimum Income Standard (MIS) study in the UK involved 12 focus groups in which members of the public from a range of social backgrounds were tasked with producing lists of items that households would need in order to reach 'an acceptable minimum standard of living'. These different groups involved pensioners, working-age adults without children and parents with children. They were complemented by and interacted with experts, including a nutritionist who helped to construct

---

[5] For example, in the UK: Mack and Lansley (1985); Niemietz (2010); Deeming (2011).

adequate diets and a heating engineer who specified home energy requirements (Davis et al. 2014).

This exercise has resulted in an agreed minimum consumption bundle that in some respects differs radically from the norm. The UK forums decided that private cars are luxuries and not necessary for a decent standard of living – citizens could use public transport and taxis instead. The parents' group, recognising that it was increasingly likely that children in social housing would be expected to share bedrooms, concluded that this was not a standard that they agreed with and felt that in general the minimum should include a bedroom for each child of school age. Necessary food expenditures were agreed to be higher than present averages, owing to the consumption of more fresh fruit and vegetables.

The implication of this research is that citizens can agree on a list of necessities at any point in time. The list will change over time in reaction to socio-technical shifts. In 2012, the group discussing the needs of households with children included a cheap second-hand car as a necessity for the first time, owing to the decline of public transport. By 2014 all groups, including the pensioners' group, regarded a computer, internet access and a cheap mobile phone subscription as a necessity. But, these apart, the definition of necessities has changed relatively slowly: the UK 2014 budgets were remarkably similar to the 2008 budgets, despite the financial crisis and recession in the meantime, reflecting a consistency in the ways that members of the public interpreted the rationales of necessity (Davis et al. 2015).

If the entire population were living on the 'decent life budget' then emissions would be lower: 37 per cent lower than actual consumption-based UK emissions in 2004 according to research by Druckman and Jackson (2010), captured in Figure 7.4. Almost every category of consumption would deliver lower emissions except for food. Reductions are particularly noticeable in transport, household energy, restaurants and hotels, and miscellaneous household goods and services. In this hypothetical scenario, total UK consumption emissions would fall from 26 tonnes CO_2e per average household to 16 tonnes.

This is a significant reduction, but it still equals 7.3 tonnes per person – well above the 2050 goal of 2 tonnes per person. In the UK and other rich countries, the emissions of the minimum consumption bundle *exceed* those of the maximum sustainable consumption bundle – indeed, they are three times greater. Studies in other countries replicate this finding: in Finland, people receiving minimum income benefits exceed ecologically sustainable lifestyles by a wide margin (Hirvilammi et al. 2013). This is to be expected and backs up the Swedish modelling in Figure 7.1: policies to recompose consumption complement but cannot replace policies

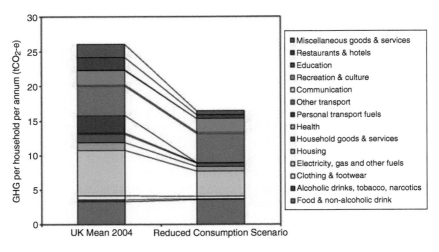

Source: Druckman and Jackson (2010, fig. 2).

*Figure 7.4   UK household GHG emissions in the reduced consumption
            scenario*

to decarbonise production. To make consumption sustainable within
existing socio-technical structures would deprive citizens of a vast range
of goods and services that they have agreed are necessary for effective
participation in modern life.

Turning from the lower to the upper bound of the consumption
corridor, can the dual strategy arrive at a consensual measure of more
sustainable maximum levels of consumption? Returning to four of the
UK focus groups and using the same methodology, one study explicitly
asked them to consider their carbon footprint and agree on what
might be 'publicly acceptable' (Druckman et al. 2011). The results were
discouraging. There was some support for reducing heating energy by
wearing more clothes at home, maintaining separate temperature zones
and switching to energy-saving appliances, and some interest in cycling
more and using public transport if it was more convenient and time-
saving. There was very little interest in reducing consumption of red
meat or eating only seasonal vegetables. The research found that people
had got used to having a high degree of choice and did not want to be
told what and what not to consume (Druckman et al. 2011). This is just
one small study in one country, but it is not optimistic that within contemporary society consumption can be consensually recomposed to any
significant extent.

More transformative approaches will clearly be needed. One approach is to use Max-Neef's (1989) framework of human scale development (HSD), first developed to enable small communities in Latin America and elsewhere to devise their own routes to a more people-centred idea of development. It involves longer participatory workshops that can question goals, behaviours, satisfiers and infrastructures more radically and over longer time spans. The HSD framework considers need satisfiers to be systematically related and interdependent and then searches for 'synergic satisfiers' that simultaneously meet different kinds of needs.

To identify these entails a 'deep learning journey' going beyond focus groups to longer-term and community-based projects (Guillen-Royo 2016). This can best take place within communities already inspired by sustainability goals, such as eco-villages and transition towns. There is growing evidence that localities are better placed to recompose consumption. Jackson and Victor (2013) provide much evidence on the way that meaningful participation encourages longer-term and joined-up thinking. Localities can bring together singular technologies to provide 'transformative networks of innovation'. Decarbonising our economies and our way of life can never be achieved by simply new technology: it requires a network of transformations that are more readily conceived and perceived on a smaller human scale, such as carbon rationing action groups (CRAGs) and transition towns in the UK (Howell 2012; Whitmarsh 2011; Steward 2012).

But there is a consistent problem in scaling up such initiatives in the face of overwhelming power imbalances and system lock-in. It is time to turn to these.

BARRIERS TO SUSTAINABLE CONSUMPTION IN PRACTICE

Theories of consumption vary on a wide spectrum from those privileging consumers as agents to those privileging corporate power and structural path-dependency.[6] I will quickly survey these before concentrating on the issues of structural power and lock-in.

At the most basic level, orthodox consumer behaviour theory condones providing information to improve the decisions of rational individuals; examples in the UK include the Energy Saving Trust and Carbon Trust providing households and firms with energy-saving information. There

---

[6] This contrast echoes Duesenberry's famous aphorism: 'Economics is all about how people make choices; sociology is all about how they don't have any choices to make' (Duesenberry 1960: 233).

is much evidence that this approach has failed: numerous studies show that many energy reduction programmes in the home would pay for themselves, yet householders are loath to take up the schemes on offer (see Chapter 6).

At a second level, behavioural economics rejects the simplicities of *homo economicus*, emphasises the importance of inertia, and explains consumer behaviour in terms of 'satisficing' – achieving an acceptable threshold of welfare rather than an optimum one. This informs policies of 'nudging': that changing the 'choice architecture' of individuals and households can influence motives, incentives and decision-making in a desirable direction without direct instruction, legislation or enforcement. Putting fruit rather than sweets at children's eye level in shops would be an example (Thaler and Sunstein 2008). But there is much evidence that 'nudging' will not be sufficient in the face of remorseless pressures to consume (Thaler and Sunstein 2008; Stoker 2009; Hodgson 2013; Room 2016).

Third, sociological theories show that behaviours cluster in sets of 'social practices', and that these are influenced by multiple factors, such as identity, belonging and social norms (Shove et al. 2012). This suggests a more variegated approach to getting people to change behaviours: recognising differences between social groups according to their understanding of environmental issues, their willingness to act and their ability to act, as applied to different types of behaviour (using a car less for short journeys, wasting less food, etc.) (Department for Environment, Food and Rural Affairs 2008). Since most researchers argue that a transition towards a sustainable society is only possible in a democratic way, ecological sustainability ultimately depends on the attitudes citizens hold towards reducing ecological impacts and hence material affluence (Soper and Emmelin 2016).

A fourth perspective stresses the corporate and capitalist framework within which consumers operate, which is heavily biased to create novel consumer goods and services and to encourage absolute increases in consumption (Seyfang and Paavola 2008). Advertising and marketing – the 'engineering of consent' (Bernays 1955) – are of clear importance in shaping consumer behaviour, not always to the benefit of consumers.[7] But these more obvious agency-based examples of corporate power rest on a less visible structural power, including control over investment and

---

[7] Two examples: one-half of poor UK households persuaded to switch energy suppliers following doorstep sales found themselves with a worse energy deal than before (Hills 2012). The vast majority of adults in Great Britain (82 per cent) have 'regretted' a purchase in the past year, amounting to between 2 per cent and 10 per cent of total consumer spending (Skelton and Allwood 2017).

Table 7.3   From basic needs to specific satisfiers

Basic needs	Social participation, health and autonomy			
Intermediate needs	Income	Nutrition	Healthcare	Relationships
Satisfier level 1	Employment	Shopping	Medical visits	Social visits
Satisfier level 2		Travel		
Satisfier level 3		Car		

the discursive power of ideas such as consumer sovereignty and economic growth (Lindblom 1977; Gough 2000, ch. 4). Fuchs et al. (2016) describe in some detail how such hidden power has delivered abundant cheap meat in modern society and thwarted attempts to regulate and limit it, despite downsides in health and emissions.

Finally, 'systems of provision' lock households into patterns of consumption that are largely outside the scope of individual choice (Jackson and Papathanasopoulou 2008). For example, many car journeys – to commute, shop, drive children to school and so on – might be caused by the spatial and social contexts within which people live and to which they must necessarily adapt. The needs framework can be adapted to provide some insights here. Need satisfiers can be viewed as a hierarchy, as for example in Mattioli's (2016) analysis of the need for use of a car (Table 7.3).

Conceiving demand in this way helps analyse the nature of the lock-in. For example, is the use of a car a necessary means to travel (level 3)? Does access to work and shopping require travel – to what extent could the internet replace work, shopping and social interaction (level 2)? To what extent does demand for travel reflect the spatial distribution of physical infrastructure (level 1)? The hierarchy demonstrates that, in defining what is needed and sustainable, attention must be paid to 'lock-in' as well as 'luxury'. This approach sheds light on some issues of justice in a climate-constrained world. Car use in a car-dependent setting may be essential for social participation and thus its denial or removal would constitute an unjust harm. Yet, if this injustice is overcome at the expense of still higher transport emissions, it will entail further degradation of the natural environment and an unjust imposition on other peoples and on future generations.

There are wide gaps between these five approaches, ranging from theories of rational consumers at one extreme to citizens confronting powerful structures at the other. What is more, the implied policies may conflict: incentives that appeal solely to self-interest may degrade intrinsic motivations such as altruism and solidarity – and thus fail to deliver their intended outcomes. Public policies will need to be aware when economic incentives conflict with value-driven goals, as Titmuss (1970) demonstrated in his comparison of giving and selling blood (Taylor-Gooby and Stoker 2011).

Jackson in *Prosperity without Growth* (2009: 101) endorses the more structural perspective:

> Individuals are at the mercy of social comparison. Institutions are given over to the pursuit of consumerism. The economy is dependent on consumption for its very survival . . . There is a rather too perfect fit between the continual consumption of novelty by households and the continuous production of novelty in firms. The restless desire of the 'empty self' is the perfect complement for the restless innovation of the entrepreneur . . . Taken together these two self-reinforcing processes are exactly what is needed to drive growth forwards . . . Nature and structure combine together here to lock us firmly into the iron cage of consumerism.

The two system-level barriers to sustainable consumption – asymmetry in power and knowledge between corporations and consumers and 'lock-in' to path-dependent structures – interact. Consumption actions result in unintended consequences that then shape future collective arrangements and future individual consumption actions. Giddens (1984) refers to this interplay between agency and structure as 'structuration'. Individual actions result via a composition effect in aggregate outcomes that have little relationship with the original intentions of individual actors. Human history is 'created by intentional activities but in a non-intentional way' (Mattioli 2016).

Yet the 'iron cage' is too strong an image. Consumption practices are affected by longer-term shifts in tastes and by socio-technical innovations and can change. For example, consumer demand for private transport is falling among young adults across the developed world for a variety of reasons, and car companies have not been able to halt that trend (Kuhnimhof et al. 2012). Similarly the entry of new market players such as Zipcar and Uber may result in reductions of car ownership and more efficient use of existing car fleets (Frenken 2017). But lock-in and luxury have been major drivers of rising emissions up to now.[8]

The conclusion thus far is that changing consumer preferences will be constrained by corporate power, system lock-in, and the interaction between the two.[9] Growing inequality also makes recomposing consumption more

---

[8]   For example, between 1968 and 2000 UK car emissions incurred for transporting children to school rose by 600 per cent, for shopping by 300 per cent, for commuting by 180 per cent and for entertainment by 120 per cent (Jackson and Papathanasopoulou 2008).

[9]   All serious studies by scholars in political science, heterodox economics, law and behavioural sciences demonstrate that consumer preferences are endogenous to socio-economic

difficult and creates dilemmas in redistributing incomes in high-carbon economies. The upshot is that recomposing consumption will require some hefty top-down state interventions. Many will be required to influence the eco-efficiency of production. But a range of novel state interventions will also be needed to recompose consumption in a fair way: a suite of 'eco-social policies' that simultaneously and explicitly pursue both equity/justice and sustainability/sufficiency goals.

ECO-SOCIAL PROPOSALS FOR POLICY INTEGRATION

In this section I propose 'eco-social policies', defined as policies that simultaneously and explicitly pursue both equity/justice and sustainability/sufficiency goals. The wellbeing and justice goals can be divided into two: those affecting the distribution of money incomes and what they can buy; and those affecting important need satisfactions and their distribution. I will discuss them in that order.

Taxing High-Carbon Luxuries

The economist Robert Frank (2011) has long argued for a progressive consumption tax, on sustainability as well as equity grounds. The polarisation of society means that luxury consumption and emissions are increasingly dominated by the global ultra-rich. Hyper-consumerism then fosters an unending expansion in wants and desires down the income and wealth scale. A progressive consumption tax would certainly contribute to curbing this Plutocene spiral. However, the tax is not as progressive as it first seems, since a progressive consumption tax equals a progressive income tax that excludes savings. Since the share of savings rises with income, this alone would benefit higher-income groups more.

But in all OECD countries except the US there exists an explicit tax on consumption – Value Added Tax (VAT) – that raises about a fifth of all tax revenues and is a major funder of social programmes. The VAT rate in most EU countries today varies between 20 per cent and 25 per cent, but in all countries there are exemptions and lower rates applied to certain goods

systems. For example: '(In the face of current levels of obesity) it is quite fantastic to suggest that everyone is choosing the optimum diet, or the diet that is preferable to what might be produced with third-party guidance' (Sunstein and Thaler 2006: 237). And 'Any welfare approach based on the presumption that individuals are always the best judges of their own interest falls at the first hurdle: many people neither understand nor accept the conclusions of the science of climate change' (Hodgson 2013).

and services. These usually include basic foodstuffs, but they vary greatly across countries, following pressures from particular industries and lobby groups.

David Fell (2016) advocates a 'smart VAT' to introduce deliberate variations in the rate – higher to discourage bad consumption and lower to encourage desirable consumption. The proposal has mainly been advocated on health and wellbeing grounds, to improve healthy eating and discourage obesity. But it could also be amended to take account of sustainability. Thus high-GHG goods that harm wellbeing would attract the highest VAT rates, while low-carbon goods that improve wellbeing would be taxed at lower or even negative rates (amounting to a subsidy).

This entails an assault on common notions of consumer sovereignty and would meet corporate, retail and consumer opposition. To secure legitimacy and public support, forms of citizenship engagement along dual strategy lines would be necessary. To decide what goods are virtuous and what harmful Fell proposes regular deliberative dialogue in focus groups informed by environmental and other experts. Their decision would then be fed up to a second stage of public decision-making to decide on the different VAT rates. Fell concedes that this stage would likely require economic expertise, such as the Office for Budget Responsibility in the UK, but to be effective smart VAT rates should vary widely between perhaps plus 25 per cent and minus 20 per cent. To publicise what is happening they should be displayed on all goods in shops and should be monitored and adjusted as necessary. Such a citizenship-based 'commitment device' (Offer 2006) would face substantial opposition from business and some individual consumers. However, there is some public opinion evidence of potential support if applied to climate-stabilising consumer behaviour (Fell 2016).

There is no reason why the emissions footprint of consumption practices should not be integrated into this process alongside health and other basic need goals. Smart VAT provides a broad framework within which specific proposals to tax high-carbon non-essentials could fit, such as a frequent-flyer levy or a global tax on business-class flight tickets (Chancel and Piketty 2015). It provides a workable method to scale up a dual strategy for sustainable consumption.

Widening Social Consumption

There are several reasons for raising the share of *public social* consumption as part of an eco-social strategy. Most generally, reducing the sphere of private consumption would reduce opportunities to compare one's consumption with other and richer groups, which is one of the drivers towards hyper-consumption. Tax-financed social consumption such as health services, social care and education is inherently redistributive: allocation according to need, risk or citizenship, not market demand, automatically serves redistributive social goals. On average, in OECD countries, public services are worth the equivalent of a huge 76 per cent of the post-tax income of the poorest group, and just 14 per cent of the richest. Public services reduce income inequality in OECD countries by an average of 20 per cent (Verbist et al. 2012; Seery 2014). In many countries this is especially strong for social services and for education, owing to the concentration of children in low-income households (Gough et al. 2011).

Second, there is evidence that public consumption is more eco-efficient than private consumption. Public services account for 12 per cent of total consumption-based emissions in the UK, or 1.8 tonnes per person – a lower rate of emissions per pound spent than private consumption, owing in part to the service nature of social services. Of the 1.8 tonnes, 0.6 tonne is emitted by health services, 0.2 tonne by all public education, one-fifth by the military, and the remainder by general public services. The majority of NHS emissions are not direct from energy use in buildings (24 per cent) but are embodied in procured goods (59 per cent) – pharmaceuticals alone account for 22 per cent.[10] A similar pattern is found in the carbon footprint of the education sector (SDC 2008).

However, publicly funded welfare states emit less carbon than privately funded alternatives. For example, the US healthcare system directly accounts for 8 per cent of emissions in the US, compared with 3 per cent of UK emissions directly stemming from the NHS. This is due both to the greater macro-efficiency and lower expenditure shares of health in the UK, and to lower emissions per pound or dollar spent, presumably as a result of better allocation of resources and procurement practices (Chung and Meltzer 2009).

Finally, this strategy could be extended to a range of 'commons' goods, such as energy and water. The 'commoning' movement (definition: 'producing and reproducing a commons') entails developing non-capitalist ways of managing resources that are central to human life, so that they are democratic, horizontal, participatory and respectful of local difference. Commoning energy is already a reality elsewhere in Europe, for example in Denmark, where imported fuel has largely been replaced with local renewables. Denmark's wind power revolution has been described as 'a grassroots, community-based initiative, underpinned by decentralised,

---

[10] In addition, the concentration of healthcare into fewer and larger hospitals generates further indirect emissions owing to patient and staff travel (Michaelson et al. 2012).

cooperative and municipal ownership alongside small-scale private ownership' (Bollier 2016). It provides an alternative route to social consumption and another potential avenue for breaking the current system of hyper-consumption.[11] But again the question arises whether it can be scaled up without state intervention – a topic returned to in Chapter 9.

Rationing: Personal Carbon Allowances

An alternative radical proposal is to ration carbon between households in some way. Various programmes of personal carbon allowances have been proposed, but all entail a cap on a country's total GHG emissions (decreasing year by year) and a division of this amount into equal annual allowances for each adult resident (Environmental Audit Committee 2008; Fawcett and Parag 2010). Children are usually recommended a lower allowance, raising issues concerning the 'emission claims' of children (Gough et al. 2011: 34–36). They could be implemented using personal carbon cards and smart metering, though the administrative difficulties should not be underestimated. In effect, a dual accounting standard and currency is developed: energy, goods and services would have both a money price and a carbon price.

Personal carbon allowances are usually assumed to be tradable, so that those who emit less carbon than the average could sell their surplus allowances and gain, while higher emitters would pay a market price for their excess. Advocates claim that a personal carbon allowance and trading (PCAT) scheme covering domestic energy, road fuel and air travel would on average be quite progressive. In addition, there is some evidence that it could generate psychological and normative motivations to encourage and sustain the kind of behavioural change that leads to emissions reduction (Fawcett and Parag 2010). However, PCAT does not avoid all issues of fairness; for example, those living in inefficient or underutilised housing, dependent on car travel, or with special needs would face difficulties. Too many exceptions to the standard allowance could undermine the scheme, but too few would result in rough justice, which could undermine public support.

It also faces other problems. It will be difficult to integrate such a downstream scheme with upstream carbon trading schemes like the ETS. The sale and purchase of excess permits would be complex to administer and likely to be less progressive than advocates claim: training every citizen to be a trader would be daunting, and richer households would likely employ

---

[11]   There are similarities here with Elinor Ostrom's (2009) 'polycentric' approach of building a strong commitment to action in small to medium governance units linked to each other through information networks and monitoring.

professional carbon managers, which most people could not afford. Derivatives and manipulation would be hard to control in the current era. It would also be costly to extend it to the wide range of consumer goods and services supplied by supermarkets. In 2012 the Tesco supermarket chain dropped its plan to label all its products with their carbon footprint, blaming the amount of work involved and other supermarkets for failing to follow its lead. For these and other reasons, the UK government in 2008 abandoned its plans to test the idea. However, a more recent series of studies considered it a suitable future strategy for delivering long-term, sustainable cuts in carbon emissions in a way that other policies cannot (Fawcett and Parag 2010).

A PRECAUTIONARY WELFARE STATE: UPSTREAM PREVENTION FOR WELLBEING

When we turn to wellbeing in the broader sense of need satisfaction, more qualitative questions arise. Welfare states have been accused of addressing immediate social problems rather than upstream causes. Yet research into health, crime, early years interventions and many other areas of social policy stresses the advantages of prevention over coping, cure, compensation or confinement (Gough 2015b). The argument is that social policy should pursue the goal of wellbeing – and counter ill-being – more directly. The case for preventive public policy is essentially twofold: normative and consequential. It is better for human wellbeing to prevent harm than to deal with its consequences. And it promises financial savings to expensive and hard-pressed welfare states.

Prevention can be defined as 'action to reduce the probability of a risk occurring'. It is built on two basic foundations: a scientific understanding of cause and effect and the possibility of prediction; and a capacity for controlled government intervention in social life (Freeman 1992, 1999). But prevention can mean many things. Coote (2012b), drawing on health policy analysis, distinguishes three levels of public interventions:

- upstream interventions: to prevent harm before it occurs, usually focusing on whole populations and systems;
- midstream interventions: to mitigate the effects of harm that has already happened, usually targeted at groups or areas considered 'at risk'; and
- downstream interventions: to cope with the consequences of harm that has not been – or cannot be – avoided.

Since I am focusing on the structural determinants of consumer behaviour, this section will concentrate on the potential of upstream interventions, with special reference to health. These do not address the immediate causes of harm but rather the 'causes of the causes' (Marmot 2005). They are concerned with the adaptation of circumstances to individual need, rather than the adaptation of individuals to circumstances. Such interventions target prevailing economic and social structures, rather than adapting individuals to them.

Advocates of upstream interventions in health range from the 1978 World Health Organization Alma Ata declaration to the 2013 World Health Organization review of health inequalities in Europe (Marmot 2013). The most radical arguments stress the 'pathogenic nature of modern social structures', for example the current 'obesogenic environment' of energy-dense food, motorised transport and sedentary lifestyles (SDC 2010). The Marmot Report proposes a broad preventive strategy that includes fair employment and good work, a healthy standard of living for all, and healthy and sustainable places and communities. But these proposals have gained little traction. Rather, what preventive health interventions there are have been biomedical (for example, statins), targeted interventions on 'at risk' groups (for example, immunisation, nutrition during pregnancy, parenting classes) and health education and lifestyle change (for example, smoking cessation) (Gough 2015b).

Could the ecological and climate threats re-galvanise the prevention agenda? Apart from adaptation, climate policy is all about prevention (though it is called 'mitigation' because past emissions cannot yet be undone). This raises the question: can prevention be extended as a more holistic strategy to social policy and economic policy? We have already noted the strong case for eco-social co-benefits. One UK study concludes that a shift in transport from driving to walking and cycling could bring about significant reductions in heart disease/stroke, breast cancer, dementia and depression. Similarly, a 30 per cent reduction in livestock production and consumption of red meat would reduce heart disease by 15 per cent (excluding effects on all other obesity-related diseases) (Haines et al. 2012).[12]

In one sense the entire welfare state can be viewed as a preventive enterprise, reducing infant mortality, raising literacy, providing public health, sanitation and hygiene more generally, providing a modicum of income

---

[12] These improvements would, other things being equal, reduce demands on health services and save money. In 2009 overweight and obesity cost the UK NHS £4.8 billion. If the incidence of obesity in all social classes had been the same as for the most affluent social class 1, the cost would have been reduced by 54 per cent, and this cost-saving would accelerate over time, amounting perhaps to 10 per cent of the NHS budget by 2025 (McPherson et al. 2009).

security, tackling social ills like discrimination, and so on. It provides buffers against incoming risks including climate change. This lies behind the consistent findings that more extensive and solidaristic welfare states not only are more egalitarian and arguably sustainable, but also yield higher levels of need satisfaction, for example in achieving shorter hours of work and a healthy work–life balance, which serves both social and environmental prevention goals (Goodin et al. 2008).

There is thus strong argument and evidence that a renewed welfare system coupled with strong climate mitigation could advance a collective agenda of precaution and prevention, achieving synergies across the social and environmental domains. But this is to ignore the third domain – the economy.

A critical problem here is that social policy has become detached from its economic policy anchor. In the early post-war decades a precautionary Keynesian management of the economy complemented welfare states. Keynes recognised both the radical uncertainty of the future in all economic management and the way that capitalism, as the engine of accumulation, exacerbates this radical uncertainty (Skidelsky 2009). But the current economic model hinders the necessary public regulatory, fiscal and mobilising initiatives that are essential for a coordinated social–environmental–economic preventive strategy. Within social policy, for example, 'contracting out' spawns and subsidises a growing network of private providers and a new form of private-interest government. Within climate change mitigation, it has seen off effective carbon taxes and new public investment in ecosystem maintenance.

A consumption-based framework, such as that advanced in this chapter, calls for precautionary policies to be pursued across all three domains including the economy. Issues such as taming finance via substantial restructuring of financial markets, active macro-economic management, and the socialisation of some sectors of investment are also part of a preventive strategy. This could buttress policies to recompose consumption, by for example developing new ways of 'valuing what matters' and imposing strict regulation of advertising especially aimed at children (Layard 2005). Today it is the absence of precautionary economic policy, coupled with its opposite – inegalitarian and debt-driven hyper-consumption – that is the main obstacle to a preventive and sustainable wellbeing state.

CONCLUSION

This chapter has studied emissions in the UK and other rich countries from a consumption perspective and explored routes towards recomposing

Table 7.4   Summary of eco-social policies in Chapter 7

	Welfare goals	CI: Ramp up eco-efficiency	C2: Recompose consumption
S1: Redistribution	Guarantee a minimum-income floor. Reduce inequality.	Compensate carbon-pricing losers. Manage inequality to enhance the effectiveness of carbon pricing.	Reduce inequality to avoid positional consumption and hyper-consumption races.
S2: Social consumption	Provide vital need satisfiers and discourage 'bads'.		Develop social consumption. Decarbonise welfare states.
S3: Social investment	Develop human and social capabilities.	Build personal and social capabilities for green new deal programmes.	Extend citizen participation and control re energy and utilities.
ES: Novel eco-social policies		ES1: Green new deal: house retrofitting and allied programmes. ES2: Social tariffs for energy and water.	ES3: Regulate advertising, especially directed at children. ES4: Tax high-carbon luxuries: smart VAT, etc. ES5: Trial personal carbon rationing. ES6: Develop upstream prevention throughout public policy.

consumption to prioritise both lower emissions and the production of necessary goods and services for wellbeing. Table 7.4 repeats Table 6.3 and adds in further roles that existing social policy and new eco-social policies could play in recomposing consumption.

The main conclusions are as follows. First, there is a critical contradiction in many high-carbon societies between securing emission reductions and ensuring an equitable distribution of real income. Simply redistributing income to low-income households could raise, rather than lower, emissions. This poses a fundamental problem for 'traditional' redistributive social policy, which follows from the high-carbon content of basic necessities, notably housing, food and travel.

Need theory and the dual strategy methodology can provide a collective, critical way of identifying what are necessities and what luxuries in a particular national or local setting. This fixes the lower bounds of a sustainable consumption corridor. I then offer suggestions as to how this could be applied to decide on the composition of a safe, upper boundary to consumption. In practice local initiatives will be crucial sites where 'synergistic satisfiers' can be identified and experimented with.

But upscaling such practices will always encounter not only powerful interests but structural determinants that 'lock-in' households to high-carbon forms of behaviour that are difficult to shift. These include dominant discourses, notably consumer sovereignty and consumer choice. New normative concepts such as human need and new social valuation techniques can help to counter these. To recompose consumption in a fair way will also require new top-down 'eco-social policies', to tax high-carbon luxuries, ration carbon at the household level, and socialise new areas of consumption.

Underpinning this, I advocate a broad public strategy of upstream prevention. This would necessarily encompass preventive action across all three domains of public policy – environmental, social and economic – both because they are linked and to ensure that each policy instrument aims for more than one target. It would face several problems, including the double burden of pursuing prevention for the future at the same time as meeting needs today (Coote and Harris 2013).

Can it work? Victor (2012) has modelled a scenario of 'selective growth', where commodities are grouped into high- and low-GHG intensity, and expenditure on the high-intensity goods and services is reduced fast to near zero. Accordingly, GDP per capita grows at the same rate as in a business-as-usual scenario but GHG emissions decline for the next 15 years before rising again, though at a slower rate. His conclusion is that selective growth – the recomposition of consumption – does offer some potential for mitigating the economic impacts of reductions in GHG emissions, but it is 'modest and short-term'. Nevertheless, this is not unimportant. It provides a consumption-oriented complement to improving the eco-efficiency of production. And it provides a transition to the more radical post-growth strategies discussed next.

Entering this territory broaches a still novel critique of conventional economics. To proclaim respect for consumer choice as the taken-for-granted foundation of policy is to respect the current factors and forces shaping preferences as either optimal or unchangeable. This is unacceptable, and has been questioned by some mainstream economists.

For example, Baumol (1990) makes a distinction between productive, unproductive and destructive labour in society. Adair Turner (2012) urges a rethinking of neoclassical economics to distinguish between the production of 'rational necessities' (productive), branded fashion goods and positional goods (unproductive), and activities causing congestion, environmental degradation and systemic crisis (destructive). As chairman of the UK Financial Services Authority Turner famously labelled some financial activities as 'socially useless' – a meaningless term in neoclassical economics. He concludes:

> We have no reason to be certain that the free flow of purely individual market-driven choice, operating under the influence of social fashion and self-interested marketing, will produce the allocation of consumption expenditures that maximise life satisfaction. And we cannot, therefore, escape the need for a continual process of political debate about whether and how we might influence the allocation in a more favourable direction. (Turner 2012: 85)

In other words, the *composition of consumption* enters political debate in an explicit way. Social policy is not just concerned with equity and distributive issues. In the age of the Anthropocene, social policy must be about changing patterns of consumption as well as redistributing incomes.

8. Post-growth, redistribution and wellbeing

> If growth automatically generated wellbeing we would now be living in
> paradise. (Latouche 2009)

> There are only two ways out of this dilemma. One is to make growth
> sustainable; the other is to make de-growth stable. (Jackson 2009: 128)

INTRODUCTION: THE CASE FOR POST-GROWTH

This chapter considers the third strategy to mitigate climate change (C3):
to reduce absolutely consumption levels so as to move to a steady-state
economy. In the immediate future reducing demand in this way applies
only to the rich world – the global North (and not necessarily to all of that);
one of the justifications is to free up some ecological space to permit devel-
opment-through-growth to proceed for a short time in the global South.

This process is usually referred to as *degrowth* or *post-growth*. I shall
use the term *post-growth* to describe the goal and *degrowth* as the route
towards it. There are many conceptions and many justifications for
degrowth (Kallis 2011, 2015) but let us follow Kallis in distinguishing
broadly between two ideas: 'less' and 'different'.[1] Degrowth refers to 'an
equitable downscaling of economic throughput' or 'a socially sustainable
and equitable reduction (and eventual stabilisation) of society's throughput' (Kallis 2011: 874). It implies a sustainable and incremental move
towards a *steady-state economy*. But this will entail radical shifts in the
basic institutions of society: not only less, but different. We cannot know
the shape of this post-growth economy, but we can discern some of its
functional prerequisites.

There are two fundamental limits to growth and thus two core arguments
for post-growth: biophysical and ethico-social (Daly 1996; Jackson 2009;
O'Neill 2016). Both have featured throughout this book.

---

[1] *Décroissance* was first coined by Andre Gorz in 1972 and was rediscovered in France in
the late 1990s (Kallis 2015).

First, growth cannot continue ad infinitum on a finite planet owing to biophysical limits, as has been argued above. To Jackson's pessimistic arithmetic can now be added Stern's (2015): given population growth and assuming that economic growth continues at its present rate (roughly trebling by 2050), then global emissions *per unit of output* must fall by a factor of seven to eight times by 2050 – as I write now only 33 years away. It seems reasonable to conclude that this acceleration of eco-efficiency is impossible. This is not to argue that such 'decoupling' is a waste of time – on the contrary it is of critical importance – just that it cannot be enough to avoid dangerous climate change. It is for this reason that in the previous chapter I considered a second route: to alter the composition of consumption towards lower-emission goods and services. This will help but it also will not be enough. It leads us to the final policy goal – an absolute cut in rich countries' consumption demand.

The second reason has already been introduced in Part I. There are sound normative arguments that economic growth does not contribute to wellbeing and may undermine it. In other words, as well as destroying Raworth's outer planetary boundary, it degrades the inner boundary and the social conditions for human wellbeing. This has been argued from the standpoint of different conceptions of wellbeing. From the perspective of happiness come arguments and evidence based on Easterlin (2001), that beyond a certain point subjective wellbeing disconnects from economic output and flatlines thereafter. The 'Big Seven' factors that affect happiness according to Layard (2011) include family relationships, work, community, friends, health and personal freedom; they include financial security but not continual income growth.

There are more robust ethico-social arguments against growth stemming from objective conceptions of wellbeing. I have argued that the satisfaction of basic human needs for participation, health and autonomy can provide a universal measure of wellbeing applicable to people now and in the future. These basic needs support wider lists of intermediate needs and need satisfiers that in turn can be measured across different social and institutional settings. Argument and evidence show that sustainable human development also disconnects from GDP after a certain level of development (see Figure 3.2 in Chapter 3).

The eudaimonic school of wellbeing supports the premise that we all have psychological needs for autonomy, competence and relatedness but that economic growth often fails to nurture and nourish them (Ryan and Sapp 2007). There is evidence that individuals whose life goals are more focused on wealth, image and fame than on relationship, personal growth and community have less self-esteem, self-actualisation and life satisfaction. More materialistic individuals are likely to be less satisfied with life, lack vitality, and suffer from anxiety, depression and addiction problems. Kasser (2011) concludes that where economic growth is a key goal of a nation universal psychological needs are undermined. Indeed this syndrome becomes self-reinforcing, as many people turn to money and possessions as a way of coping with distress (Koch 2013).

Furthermore, the pursuit of growth has often been at the cost of growing inequality of wealth and income. In the present period growth in many rich countries is worsening the distribution of both need satisfaction and wellbeing. An equitable and just economy cannot be guaranteed by market forces nor by the pursuit of growth. Another way that growth undermines wellbeing is through the commodification of more and more aspects of life. This is incompatible with social practices and relationships that are independently conducive to human wellbeing (O'Neill 2016). In Marxist terms, the unending pursuit of exchange value can degrade the production of use values outside the market. Relentless pursuit of growth conflicts with the goals of sustainable wellbeing and prosperity.

POST-GROWTH BY DESIGN: TOWARDS A STEADY-STATE ECONOMY

If these arguments convince, then what type of economy and society do they lead us to envisage? Herman Daly has done much to answer this question with his concept of a 'steady-state economy' (SSE) (Daly 1977, 1996). In a steady state the aggregate 'throughput' is constant and at a level that is sustainable for a long future. Throughput is 'the flow beginning with raw material inputs, followed by their conversion into commodities, and finally into waste outputs' (Daly 1996: 28). This flow must lie within the regenerative and assimilative capacities of the ecosystem. It implies a constant stock of physical capital capable of being maintained by a low rate of material throughput.

A sustainable level of throughput is not achievable by market forces, just as an equitable economy is not. At the most abstract level of welfare economic theory, 'Pareto efficiency' indicates a state where no one can be made better off without someone else being made worse off. But we know that this is compatible with any distribution of income and wealth however unequal; it simply takes the distribution for granted. Daly's insight is that the same applies to the scale of resource input. So an economy can be 'Pareto efficient' however unequal or unsustainable it is (Daly 1996; cf. Pirgmaier 2017).

Thus the three values associated with the three domains of life set out in the Introduction to this book – allocation (efficiency), distribution (justice) and scale (sustainability) – can conflict. Markets are primarily seen to improve allocative efficiency, which is implicitly regarded as the prime goal in neoclassical economics. But when justice and sustainability are prioritised the primacy changes. A more discriminating principle for balancing the present and the future would be that the basic needs of the present should always take precedence over the basic needs of the future but that the basic needs of the future should take precedence over the extravagant luxury of the present (Daly 1996: 36). This takes us back to the premise of this book.

A steady-state economy is not a stagnant economy. There would be continual change with some sectors expanding and others contracting. Nor does it necessarily equate with zero growth of GDP (cf. van den Bergh 2011). Yet research into economic throughput is limited, partly owing to complexities in operationalising the notion. Thus several researchers investigating the macro-economics of such an economy revert to GDP (Victor 2008; Jackson 2009). The implicit assumption is then that zero growth of GDP provides an acceptable proxy for an SSE. For example, Victor's (2008) model of the Canadian economy over a 30-year period from 2005 to 2035 analyses alternative scenarios to investigate whether slower growth can be achieved without worsening social outcomes, indeed whether it can be combined with improved social outcomes and reduced emissions.

Victor finds that this is possible in an economy with the following features: average paid working time is cut by 14 per cent over the 30-year period, the share of both investment and government social spending rises – the latter to provide a minimum-income floor and invest in social services – and a very high carbon tax is imposed (C\$200 per tonne). With these policies in place growth of GDP slows down from 1.8 per cent per annum to near zero. Yet at the same time unemployment and poverty are halved, emissions are cut by 20 per cent (quite inadequate, but a start), and government debt is cut by 75 per cent – another precautionary measure reducing risks to future generations (Victor 2008). However, his subsequent modelling is more pessimistic, making a sharper distinction between zero growth and degrowth or negative growth. Stabilisation at a high level of consumption will not be enough, and to achieve degrowth requires far more drastic measures (Victor 2012). No doubt much more work is needed to illuminate the parameters of a degrowth economy.

Ecological Investment

The balance between consumption and investment has to change in a new ecological macro-economics, for two reasons (Jackson 2009). First, the

Table 8.1   *Dimensions of investment in a post-growth economy*

	Business sector: commercial rate of return	Public sector: quasi-commercial rate of return	Public sector: social rate of return
Energy efficiency	x	x	
Renewable supply	x	x	
Climate adaptation		x	x
Ecosystem maintenance		x	x

driving role of consumption in current economic growth must be curbed. Second, the economy and society need to be reshaped and directed along a different pathway, but to do this requires capital investment. There will be an urgent need to invest in renewable energy, energy networks, transport, communications, transformed cities and buildings, retrofitting housing, the preservation and enhancement of natural resources, and adaptation to climate change. Unless other forms of investment are reduced to an equal extent, this would mean reversing the decline in investment shares across the OECD over the past three decades.

But the very role and nature of investment also has to change. Table 8.1 presents a modified version of the necessary categories of investment described by Jackson (2009: 213).

In a post-growth economy the private business sector could continue to play a role in providing renewables and boosting energy efficiency, but all sectors of eco-investment will require some public involvement. Moreover, when looking at very long-term and difficult-to-quantify benefits, such as climate adaptation and 'ecosystem maintenance', the public sector will play a major role. Only it can develop different criteria for calculating return on such investment, emphasising social and long-term or very long-term goals. This will mean revisiting the concepts of profitability, productivity and growth.

This conflicts with conventional economics, where these sorts of investments are regarded as 'unproductive' and not usually included in the 'production function'. Whatever their importance for sustainability, they 'soak up' income produced elsewhere in the economy (Jackson 2009). This recalls earlier debates on the political economy of welfare states. For example, according to Bacon and Eltis (1976), the economy can be divided into two sectors: the marketed sector; and the non-marketed sector comprising all social services and other government activities such as law and order, and defence. Their central argument was that the marketed sector had to supply the total commodity needs of the nation – consumption, investment and exports, including commodities for the growing numbers of public sector workers in health, education and care. A falling number of productive workers had to supply a rising number of unproductive workers.

In response to this I have argued that their analysis ignores the return flow of de-commodified services or use values provided by government that all households enjoy (Gough 1979). Instead of purchasing health and education as commodities in the private market, households in a welfare state pay taxes and receive the services free or heavily subsidised. What private (and public sector) employees lose in taxation they more than gain in services for which they would otherwise have to pay. The same argument applies to Jackson's 'ecosystem maintenance': the costs of climate change will ultimately, and sooner rather than later, impact on society including the market economy. The necessary eco-investment and subsequent adaptation costs must be paid by society as a whole, whether by higher taxation or subsidised low-interest loans.

One can generalise this dilemma as follows. Collective social services are essential for social cohesion and wellbeing; and collective investment in ecosystem maintenance is essential for environmental and social sustainability. Both sets of activities are necessary for the long-term sustainability of society; they are *reproductive* activities. Yet they are often regarded as 'unproductive' within capitalist economies because they are not independently funded from flows of consumer spending.

On the other hand some labour in modern societies will be *unreproductive*: labour that does not contribute to, or actively undermines, the longer-term sustainability of society. Mention has been made above of 'socially useless' labour in the financial sector, such as some dealers in complex financial instruments in the City of London. Others would include many workers employed as 'guard labour' in prisons, private security firms and so on. Bowles (2012) calculates that the entire 'garrison economy' employed 26 per cent of the US labour force in 2002 – a tremendous burden unavailable for social reproduction. Categories of reproductive and unreproductive labour require new forms of social valuation for their contribution and burdens to be recognised.

Population

A steady-state economy must ultimately be predicated on stable population levels. And, since (other things being equal) more people imply more greenhouse gas emissions, climate change raises anew the question of appropriate population size. Jonathon Porritt, former Chair of the UK Sustainable Development Commission, for example, has warned that

Britain must drastically reduce its population if it is to build a sustainable society. And the Optimum Population Trust advocates a goal of halving the UK's present size to 30 million people.

Population policy has virtually disappeared from public discourse since the 1970s, when it animated discussion on the limits to growth (Ehrlich and Holdren 1971; Meadows et al. 1972). Just 40 years later world population passed the 7 billion mark, and it is predicted to rise to at least 9.5 billion by 2050. Yet despite the impact of continuous population growth on all aspects of environmental sustainability – not just climate change – it only intermittently figures in degrowth discourse. Coole (2013) contends that this reluctance stems from several 'silencing discourses'. One of these is 'population-shaming', which treats limiting population as a front for more despicable motives to limit certain categories of people and to blame the fecund poor for climate change. Another is 'population fatalism', which treats the challenge of 9 billion people as a factor that cannot be challenged or changed.

Another discourse, 'population declinism', increasingly influential in the North, begins from the fact that fertility rates are well below replacement levels in many European countries. This stokes anxiety over unsustainable welfare costs and falling growth rates and paradoxically calls for increased immigration as the surest route to maintain population size. Unfortunately this is a short-term fix, since immigrants too will age, prompting calls for still further immigration. The worry that a declining or stable population will negatively affect economic growth is explicable from both conventional and green growth perspectives, but makes no sense in the context of degrowth, where population stability should be welcomed and praised.[2]

In practical terms it is true that the bulge in the numbers of young women of child-bearing age means that the global number of future births will grow over the next two to three decades. But the policies to slow this down are well known. The two critical drivers of fertility are girls' education and the employment of women (Anand and Sen 2000), which are also major components of need satisfaction and human wellbeing. Policies to promote these, and to remove obstacles to women's control over their fertility, are fundamental and desirable eco-social policies.

This still leaves major socio-economic issues for further debate, including the global model of neoliberalism and free movement of goods, capital and people. Several green economists and parties call for a move away from such an economic model towards a more localised system of

---

[2] Other things being equal, immigration from poor, low-carbon countries to rich, high-carbon countries will also increase global emissions.

provisioning. Though the definition of 'local' is open to debate, a move in this direction would necessarily imply more national economic regulation. It implies a shift from a global model of free movement of people, capital and goods to a UN model of a family of nations, committed to some form of 'progressive protectionism' plus much greater financial flows to promote sustainable development in the South (Hines 2017).

RETHINKING POST-GROWTH WELFARE

Economic growth has sustained the welfare state as we know it in at least two ways. First, rising GDP has delivered rising tax revenues to fund the inexorable rise in welfare demands and costs. Second, growth facilitates redistribution from the better off to the worse off without the better off losing out in absolute terms (Crosland 1956). Economic growth supplies economic resources and political legitimacy to the welfare state. It is no coincidence that the post-war welfare state emerged simultaneously with the modern tax state and growth state.

However, this fundamental justification for welfare capitalism has been eroded over the past third of a century. Since around 1980 modern capitalism in much of the North has shifted from growth plus redistribution to growth plus rising inequality to much lower growth plus still high inequality. Are we, after the financial crisis, already entering unplanned 'post-growth'? If so this shift in the underlying political economy might be seen in some ways to prepare the ground for post-growth by design. But what are the further implications of 'smaller by design'? I consider just two here: a) managing the falling share of wages in the national income; and b) funding the accumulated entitlements of social security and public healthcare schemes.

Piketty's 'second fundamental law of capitalism' claims that slower growth increases the ratio of the stock of wealth or capital (he does not make a distinction between the two) to national income. 'In a quasi-stagnant society, wealth accumulated in the past will inevitably acquire disproportionate importance . . . The return to a structurally high capital/income ratio in the 21st century . . . can therefore be explained by the return to a slow-growth regime' (2014: 166). This finding is clearly portentous if the policy goal is deliberately to reduce growth in rich countries to zero. It puts redistribution, wellbeing and the social domain in the centre ground.

Second, the future of social protection is threatened when ageing populations and falling fertility increase the share of total social expenditure in GDP. This is a special problem in the Eurozone, where social protection expenditures have doubled as a share of GDP from around 15 per cent in the late 1950s to 30 per cent of GDP in the 2010s. Looking ahead, the European Commission expects them to reach almost 35 per cent of GDP by 2050. Yet falling growth worsens these scenarios. One study forecasts that if growth in France fell by one percentage point below standard forecasts social protection spending would climb to 50 per cent of GDP in 2050 (Demailly et al. 2013).

These are worrying dilemmas. The conventional solutions appear to be to raise productivity growth, or step up contribution rates, or reduce pension and health service levels relative to wages. Yet for some critical economists productivity growth is the problem not the solution (Jackson 2009). And we will argue below that for climate reasons productivity growth – an increase in output per hour worked – must be used to reduce paid work time.

Post-growth poses fundamental challenges for the pursuit of wellbeing. It requires rethinking at three levels: the redistributive 'welfare state' at the national level, the role of the core economy at the local level, and the role of the 'social commons' within the public sphere.

1. Top-Down Solutions: Sharing Income and Wealth

We have already rehearsed in previous chapters the arguments for greater equality: economic, social and environmental. The centrality of human needs and sufficiency also invokes equality. When the cake shrinks its distribution becomes critical. Wellbeing can no longer be conceived or pursued as a race towards an ever-receding finishing tape. Rather it is a collective endeavour between us, now.

A stationary welfare state would thus need to promote more equality. One route to achieve that would be a minimum-income floor and a maximum ceiling (Daly 2007). Most welfare states, even today, incorporate some minimum-income floor together with citizenship-based social rights, notably to education and healthcare (the US remains a notable exception). However, increasing numbers are falling though holes in the net and face demeaning investigations and treatment when claiming benefits. The merits of universal basic income as a solution to this problem are discussed in detail below.

At the other end of the income scale a no-growth economy would need to rein in limitless wants to provide for basic needs within a constricting resource envelope. This would require ways of capping excessive incomes and wealth, and the case has been made for a *maximum* income (Daly 2007). There are issues to be faced about defining a maximum income[3] and

---

[3] One method, advanced by Medeiros (2006) and Concialdi (2017), has much to commend it since it links together the minimum and the maximum. It comprises two

devising a suite of policies that are desirable and feasible. Weisbrod and Hansen (1968) suggest a ratio of top to bottom incomes of 6 or 7 to 1. A Swiss referendum in 2013 proposing a 12:1 ratio was lost 65 per cent to 35 per cent, but 35 per cent suggests considerable support for the idea.

Such bounds on the distribution of incomes would challenge market valuation of different jobs. Orthodox economic theory claims that, under certain assumptions, a person's income will reflect his or her 'marginal productivity'. However, the assumptions are many and the observed disparities in incomes defy any rational justification. A post-growth society would need to undertake an alternative *social* valuation of occupations and professions, such as that pioneered by the New Economics Foundation (Steed and Kersley 2009). To demonstrate legitimacy the method would again need to utilise some variant of the dual strategy: groups of citizens informed by expert analysis of social and environmental impacts and social responsibility. The NEF report found large discrepancies between the social and market valuation of occupations, suggesting elements of an inverse law, where socially beneficial jobs are paid least and the least beneficial or most harmful are paid most.[4]

The only coherent solution to both challenges would be to spread the ownership of *wealth* and capital so that all citizens have rights to part of that dividend, but in a collective rather than an individualised way. A reallocation of property rights would give everyone a stake in capital and a non-labour source of income. This would require on the one hand steep taxation of inheritance, land and capital transfers and on the other hand building up the state's store of public capital. A post-growth economy, characterised by a rising share of capital in income, falling hours of work, and problems over the sustainability of social insurance, will strengthen the case for spreading the ownership of wealth.[5]

In many respects this runs completely counter to recent trends towards the privatisation of national assets. For example the UK, admittedly the

---

stages: first, calculate the total minimum-income gap in a country, that is, the total sum of money needed to bring everyone up to the minimum-income standard; second, starting at the top of the income distribution, calculate how far down we would need to move to accumulate that sum of money. That income would then determine the maximum income in a country.

[4] For example, childcare workers were estimated to generate between £7 and £9.50 benefits to society for every £1 they are paid, and waste recycling workers £12. Against this are occupations that actively destroy social value by reducing taxes paid by the rich (tax accountants destroyed £47 of social value for every £1 paid) or by generating gross economic and social insecurity (investment bankers destroyed £7 of social value for every £1 paid).

[5] An earlier proposal made by the Nobel economist James Meade (1964) advocated a 'property-owning democracy' as one way of avoiding what he called the 'hideous outlook' of millionaire property owners and an immiserised proletariat caused by technology shifts and automation. A similar concern is evident today fuelled by the IT revolution.

OECD nation most committed to privatisation, has run down all forms of public wealth from an amount equivalent to three-quarters of national income in the late 1970s to *minus* one-quarter by 2013 (Höpner et al. 2014; Lansley 2016). Many countries run sovereign wealth funds, usually financed from trade surpluses and profits and dividends from oil wealth, for example in the Middle East, China and some US states. A few of these use the proceeds to pursue social or welfare objectives. For example, the Norwegian fund is designed to spread the benefits of its oil wealth across future generations by purchasing financial assets to compensate for the dwindling oil assets.[6] The Alaska Permanent Fund invests one-quarter of the state's oil revenues and pays out an annual social dividend to every citizen in the state, including children; up to now the amount paid has varied between $900 and $3300 according to the price of oil (Lansley 2016).

Perhaps the most radical reform plan along these lines was the Meidner plan in Sweden. This proposed an annual levy on the capital of the wealthy by forcing companies with more than 50 employees to issue new shares every year equivalent to 20 per cent of their profits. These shares were to be given to a set of 'wage-earner funds' representing trade unions and local authorities. Over time these funds would be reinvested and yield dividends that would finance future social expenditure, including pensions. The share levies would provide all members with a claim on future output. In contrast to private funded pensions which undermine social solidarity, the Meidner plan would also have fostered a more collective intergenerational solidarity. Following opposition by large corporations and parts of the public the scheme was diluted before being wound up in 1992 (Blackburn 2013).

The only sure ways to provide security of income in a post-growth society will be to tax or socialise a substantial part of private wealth. Post-growth thus challenges capitalism in (at least) two ways. First, it directly confronts its innate drive to accumulate capital. 'There is no way out of the dilemma between the capitalist imperative to accumulate and the limits that nature sets' (Altvater, cited in Demaria et al. 2013). Second, to avoid a devastating impact on material living standards it will be necessary to question and partially dismantle a defining feature of capitalism – the private ownership of the means of production.

---

[6]   In contrast to the UK, which spent the entire proceeds on tax cuts, mainly for the better off. Had the UK accumulated such a fund it would now be worth at a minimum £350 billion (Atkinson 2015).

2.   At Ground Level: Rebuilding the Core Economy

A post-growth society must extend beyond a (transformed) state and a (truncated and transformed) market sector. It will necessarily draw on and foster the uncommodified resources that remain in modern societies. The whole ensemble of resources and activities that sustain human life has been called the 'core economy' or the 'social economy' (Coote 2011; Barry 2012). The core economy comprises the resources embedded in the everyday lives of every individual (time, wisdom, experience, energy, knowledge, skills) and the relationships among them (love, empathy, responsibility, care, reciprocity, teaching and learning). They are largely uncommodified and unpriced yet they have value and can be exchanged. The core economy can flourish or be damaged by the wider social environment: some neighbourhoods are alive with activities that enrich and strengthen social connections, while others are beset by deprivation, division or distrust.

Proposals to foster the core economy include devolving power and encouraging local action and building capacity, especially among deprived neighbourhoods. *Time* is a key resource, and policies to redistribute time can help; these include reducing working time and policies to diminish the gendered nature of caring. Upstream prevention, discussed earlier, will also help strengthen the core economy by building autonomous capacities that facilitate participation and action.

This bottom-up approach also generates a particular way of relating to state social services: 'co-production' (Ostrom 1996; Cahn 2000; Coote 2015). Co-production essentially means delivering public services in an equal and reciprocal relationship between professionals, people using services, their families and their neighbours. It is advocated for both intrinsic and extrinsic reasons. It is claimed to strengthen the capabilities of people, especially those with fewer financial resources, and to strengthen peer support networks. In turn this fosters better support and greater efficiency than top-down delivery. By changing the way people think about and act upon 'needs', this approach promises more resources, better outcomes, and a diminishing volume of demand for services. The idea has been taken up in some departments of UK government and has been developed in some detail by the New Economics Foundation. The guiding idea is that co-production should be the default procedure in all public services. It holds special promise as a component of social policy in a post-growth society.

3. Rebuilding the Social Commons

Between the state and the core economy lies the public sphere: a discursive space where individuals and groups associate to discuss matters of mutual interest and, where possible, to reach a common judgement about them (Habermas 1989). It has already featured in our discussions of human needs and how to secure agreement over need satisfiers (Chapters 2, 7).

A flourishing public sphere is a vital precondition for 'benign post-growth' (Dobson and Bell 2006). It requires the cultivation of 'green citizenship' to overcome several problems already encountered with economic incentives and self-centred behaviour and to encourage alternatives. The strong sustainability necessary to forestall dangerous climate change will require a more ethical sense of mutuality, collective interest and the common good. This in turn entails public action and collective engagement, such as carbon reduction action groups and transition towns (Whitmarsh 2011). The public sphere plays a central role in Nobel prize-winner Elinor Ostrom's (2009) 'polycentric' approach to tackling climate change: building a strong commitment to action in small to medium governance units linked through information networks and monitoring. The public sphere is 'where people learn non-contractual habits' (Dobson 2011). Jackson and Victor (2013) provide a convincing argument with numerous examples as to why thriving communities are the basis for shared prosperity.

These ideas overlap with forms of commons, online sharing, eco-villages, cooperatives, back-to-landers and local currencies. Examples of the 'commoning' movement include the spread of local energy coops in Denmark, movements to de-privatise local energy generation in Germany, and the anomaly of Eigg, a tiny island in Scotland owned collectively by its inhabitants and entirely supplied by renewable electricity. There are growing calls to generalise these initiatives by coupling decentralised ownership and control over energy with new forms of energy democracy (Trade Unions for Energy Democracy 2012).

In its simplest meaning, the commons refer to natural resources that should be accessible to all members of society, such as air, water, mountain, land and forests, features of the ecosphere that cannot or should not be privatised or commodified.[7] But the idea of 'commons' can be extended from natural to

---

[7] 'The commons are not concessions. They are resources that belong to the people as a matter of life necessity. Everybody has a right of an equal share of the commons and must be empowered by law to claim equal and direct access to it. Everybody has equal responsibility to the commons and shares a direct responsibility to transfer its wealth to future generations' (Ugo Mattei, http://wealthofthecommons.org/essay/first-thoughts-phenomenology-commons).

social resources. It is applied to cultural goods, such as knowledge, cultural heritage, literature, music – and the digital commons like software. But one can go further still to include all the social services and activities funded and provided collectively through our public institutions and community-based organisations. This is the 'social commons' that does, or should do, so much to meet our common needs and protect us against risk (Mestrum 2016). The idea of the commons helps to reimagine the welfare state and integrate the environmental and social domains, the subject of this book.

Much of this is clearly speculative. But we can assert with confidence that sustainable and democratic eco-social policy will require some combination of, and balance between, a transformed state, a revived core economy and a flourishing eco-social commons.

The chapter concludes by surveying at some length two popular eco-social strategies to begin the transition to post-growth: a universal basic income and reduced working time. I conclude that the former is neither feasible nor desirable, but that reducing paid work time offers a direct and effective route to just post-growth.

THE ILLUSIONS OF UNIVERSAL BASIC INCOME

A growing number of social movements, such as the Basic Income Earth Network, are calling for a universal unconditional basic income (UBI) for all citizens or residents of a territory. Not all of these are motivated by environmental concerns, but it is notable that UBI has been promoted since the early 1970s by green parties and degrowth advocates (Schneider et al. 2010). This section briefly reviews the case for and against a UBI and how it may or may not fit with degrowth scenarios.

A UBI is an unconditional payment made regularly to every individual as a right of citizenship. It would aim to provide 'a basic amount on which every citizen can survive', usually excluding highly variable expenses such as housing or extra costs for disability living. A UBI would ensure a minimum income for all citizens as of right rather than in response to their particular circumstances, or as a benefit earned by paid work and social contributions. There are several arguments made in its favour with a link to sustainability and post-growth. First, it would provide more freedom of choice over citizens' life courses; it would promote a better work–life balance, enhance gender equality and expand choices between paid and unpaid work. It might enable more people to contribute to the 'core economy'. Second, it is claimed it will provide a solution to the labour market disruption that, it is assumed, will be caused by automation and will address intensifying precarity, especially of young people. Third, it is 'minimally presumptuous': a UBI entails no official enquiries into a person's activities or household arrangements, in sharp contrast to assistance benefits in many welfare systems (Goodin 1992). Consequently, it could reduce division and stigma and enhance social solidarity (Torry 2016).

Yet all extant proposals for UBI envisage a supplemental benefit well below national poverty lines and minimum wages, let alone the decent living standard promulgated in Chapter 7.[8] This is because of the arithmetic: a full UBI would absorb a tremendous share of national income and require big increases in taxation. As the economist Tony Atkinson explains, if a 20 per cent tax rate is needed to pay for all other government spending, then to provide a basic income set at only 50 per cent of average income would require additional taxation of 50 per cent of average income – a total tax take of 70 per cent.

The recent upsurge in interest in UBI is partly due to a renewed concern that digital automation will destroy large numbers of jobs and intensify the precarity of life for millions of people. Yet this vision of a surge in labour productivity is precisely the problem according to Jackson (2009) and other advocates of post-growth. An equally likely future scenario is one of low economic growth or 'secular stagnation'.[9]

There are powerful normative arguments against UBI stemming from a wider conception of wellbeing. Liberal and libertarian arguments for UBI stress its role in achieving real freedom for all by enabling all to choose freely between work and leisure (Van Parijs 1995). But, from a human need perspective, participation in productive and reproductive activities, as well as contributing to collective welfare, is a crucial component of self-respect, contributes to cognitive development and provides a site for purposeful socialisation. These benefits from participating in socially significant activities are sidelined when the argument is framed as a simple calculus of choice between work and leisure. It is for this reason that most participants in experimental research value both need and desert, and favour the provision of a guaranteed minimum income but not unconditionally (Frohlich and Oppenheimer 1992; Gough 2000, ch. 9; Brock 2009, ch. 3). Partly in response to this the economist Tony Atkinson (2015) has

---

8 Most current proposals are not for a full UBI at all, but for variants which undermine its core rationale. These include 'partial' basic income, payable to certain groups in society, such as children (there is nothing new in this), or 'supplemental' UBI that provides a small universal benefit alongside existing social security benefits (Piachaud 2016; cf. Reed and Lansley 2016).

9 Some economists argue that productivity will continue to fall, for several reasons: the shift to low-productivity services will continue, though at a reduced rate, and the debt overhang will remain (Gordon 2012; Demailly et al. 2013).

suggested a universal participation income.[10] This is much closer in spirit to the concept of active wellbeing argued in this book. On the other hand it resurrects an administrative challenge absent from pure UBI, that of deciding what and how much constitutes participation.

Another serious problem within a post-growth scenario is that UBI focuses on money income, whereas we have demonstrated above that collective public provision of several need satisfiers is superior on grounds of efficiency, equity and sustainability.

The coupling of UBI with degrowth reveals a dilemma at its heart: an expanded and fiscally more demanding state is superimposed on a shrinking economy. This is of course a general dilemma, facing existing welfare states and publicly guided climate mitigation states. The pressing need will be for a radical investment strategy: to decarbonise production and consumption, for climate adaptation programmes, for social consumption rather than individualised consumption. If it diverts attention from collective goods, services and investment, a partial UBI could end up re-commodifying elements of existing welfare states (which is precisely the aim of its neocon advocates such as Charles Murray 2016). A needs-based approach supports a more mixed package of policies. Policy analysis shows that there is no one-to-one relationship between policy instruments and outcomes, and that one-size-fits-all policies rarely succeed (Gough 2000, ch. 9).

My conclusion is that UBI cannot provide a realistic transition strategy from the present to a post-growth society. UBI requires a one-off, top-down transformation: the abolition of numerous entitlements, acquired social insurance benefits, and tax reliefs. The idea of a citizen's income resembles a 'silver bullet', a transformative shift that could distract attention from the complex underlying causes of inequalities, ill-health and social conflict, for example. These require 'upstream' systemic changes, rather than a single downstream intervention.

THE MERITS OF REDUCED WORKING TIME

Can reducing time spent in paid work provide an alternative route to post-growth? Increases in productivity can always be taken out either in increased incomes and consumption, or increased 'leisure', or some com-

---

[10]   Participation can be broadly defined as making a social contribution – for example by full or part-time waged employment or self-employment, by education, training or active job search, by home care for children or the elderly or disabled, or by regular voluntary work in a recognised association, or a portfolio of activities equalling around 35 hours per week.

bination of the two. Shifting the balance towards more disposable time and away from the consumption treadmill makes intuitive sense, both as a characteristic of a sustainable economy and as a means of getting from here to there. It is also a prime example of an eco-social policy pursuing two distinct goals. For these reasons this section surveys at some length motivations for reduced working time (RWT), arguments and evidence for its effectiveness and policies to achieve it. It concludes by looking at distributional effects and asking who 'pays' for RWT.

RWT has been advocated in the past on social and economic rather than environmental grounds (Hayden 2013). The *economic* grounds include that it permits more flexible use of the labour force and can improve productivity and company competitiveness. The *social* grounds include beneficial effects on the health of workers and more general improvement in wellbeing and quality of life. Work-sharing has often been advocated as a way to reduce unemployment, which has been a major factor undermining both objective and subjective wellbeing. RWT is also advocated as a means to promote more gender equality in the labour market and to improve work–life balance.

A still wider case can be made for enhancing discretionary time in affluent economies as a contribution to enhancing autonomy and wellbeing. Robert Goodin et al. (2008) conceive 'discretionary time' as time over which you have autonomous control, defined as time left over after the necessary time spent in wage labour, unpaid household labour and personal care. Here, and in related work (Burchardt 2008; Coote and Franklin 2013), time is situated in a more sociological context, far removed from the simple work–leisure trade-off of economists. People have responsibilities strongly shaped by social norms, such as caring for children and dependent relatives, which make claims on their time. The remaining discretionary time may be used in many ways, including participating in a wide range of activities within the public sphere that go well beyond leisure (Coote and Franklin 2013). There are numerous other ways that RWT can contribute to basic needs for health, autonomy and participation discussed below.

The *environmental* case for RWT emerged late in the day, first in the 1970s through such pioneers as Andre Gorz (1982) and later in the 1990s through the work of Schor (1991, 2005), Latouche (2009), Victor (2008) and others. There are two distinct environmental arguments. Trading pay increments for more disposable time limits the rise in consumption and thus emissions that would otherwise take place; it would weaken the 'work and spend' cycle, which locks employees into a trajectory of fixed hours and rising consumption (the *scale* effect). Second, working shorter hours is likely to change the time and expenditure budgets of households in a lower-carbon direction; for examples, households may shift towards more time-intensive and less carbon-intensive ways of travelling, shopping, preparing food, repairing goods and so on (the *composition* effect). The composition effect is part of the case for recomposition developed in the preceding chapter of this book. But the scale effect is new: it acts to brake the absolute level of consumption in a household and a nation.

There is mounting evidence to support the environmental and climate benefits of RWT. Schor (2005) finds a significant correlation across the OECD between national ecological footprints and average hours per employee. Subsequent multivariate studies support this and find a big effect: a 1 per cent reduction of work hours per employee reduces energy, environmental and carbon footprints by between 1.2 per cent and 1.3 per cent (e.g. Knight et al. 2013 on GHG consumption-based emissions).[11] One study using this approach again concludes that reduced work time is associated with reduced energy use and GHG emissions, though the decrement is lower at 0.8 per cent for each 1 per cent reduction in work time (Nässén and Larsson, 2015).[12]

Importantly for the argument in this chapter, all studies find the scale effect to be larger than the composition effect, usually much larger. The household-level study finds changes in time use to be almost carbon-neutral when holding other factors constant; it is the reduction in incomes and consumption expenditures following on from RWT that reduces emissions. Cross-national studies find both factors significant, but the scale effect roughly twice as important (Knight et al. 2013).

Work-Time Policies and Their Distributional Aspects

Given this evidence, what policies have brought about, and could bring about, shorter working hours? There are many, but they can be divided into two groups: *collective* and *individualised*.

The most general RWT policy is to declare a shorter standard work week. The norm remains 40 hours a week across many countries today, but several European countries have broken through this barrier, as a result of either legislation or negotiation. France pioneered laws in 1998 and 2000 to cut the standard work week from 40 hours to 35 hours, though for

---

[11]   There are fewer studies at the household level, since these require imputing carbon impacts to a whole range of personal activities from sleeping to travel, from domestic work to recreation, and then drawing on time-use studies to calculate the overall carbon impact of reduced time in paid work (Druckman et al. 2012).

[12]   This figure is closer to the estimates for cross-country income elasticities of CO_2 emissions given in Chapter 3.

social and economic not environmental reasons. In practice an employee's work time could be cut in a variety of ways on a weekly or annual basis. Though much criticised in the Anglosphere and weakened by incentives to work overtime, the principle of a 35-hour standard has not been rescinded in France. Other countries have cut working time via collective negotiations, including the Netherlands, Denmark, Norway, Belgium and, in some sectors, Germany. The EU also has a 48-hour maximum work-week directive, though the UK has an opt-out. The recession following the 2008 crisis has boosted variants of RWT in several countries, notably Germany (Hayden 2013).

Another collective route to RWT consists of legislated or negotiated paid holiday or rights to paid annual leave. Almost all affluent countries have such rights except the US. The 1993 EU Working Time Directive stipulates a minimum of four weeks' paid leave, and several member states have more; the average in the US is nine days a year. Similarly, the number of public holidays varies widely and affects average annual hours of work (Hayden 2013).

An alternative approach gives workers *individualised* rights to choose to work shorter hours. It is remarkable that so many workers when signing a job contract are denied any choice over the hours they work (Skidelsky and Skidelsky 2013). Since 2000 employees in the Netherlands have had the right to reduce their hours of work, while part-timers can increase theirs. Several other European countries give parents with young children such rights. Denmark has pioneered a system of sabbatical leave for employees – for education, for childcare or to undertake a change of direction. Belgium also has a time credit system enabling sabbaticals with an income guarantee. Finally, some policies form a bridge between individual entitlement and collective provision, notably paid parental leave schemes to give parents space to look after children and to encourage gender equality. Here the Scandinavian countries have the most generous benefits.

There is a sharp divide between the US and the EU in average hours worked and in almost every form of RWT policy. Since 1975, when they had similar hours of work, the US has reduced average hours by 4 per cent and Germany by 22 per cent. All other things being equal, Germany has deployed its productivity dividend in a less environmentally harmful way than the United States.

The distributional impact of RWT raises interesting questions, even putting aside the numerous non-monetary benefits that follow from the policy. The fundamental starting point is productivity growth and how its rewards are taken. The assumption in all the studies cited above is that incomes from employment track the decline in hours of work. In other words, a growing share of productivity growth would be taken in the form of time rather than money.[13]

Does this strategy imply that the burden of RWT falls on workers (Kallis et al. 2013)? Trade unions in the 1970s argued for RWT with no loss of hourly pay in order to prevent the benefits of rising productivity accruing to capital owners. The more recent examples of negotiated RWT demonstrate that some unions value more the non-monetary benefits of discretionary time, including reduced unemployment. Yet it is mainly higher-paid employees who take up optional time reduction schemes (Pullinger 2013). Low-paid workers have little or no scope to cut their earnings and would suffer from any equal distribution of RWT. The distributional aspect of RWT should be recognised, especially in an era when the share of labour in national income is falling and inequality of all incomes is rising.[14]

Some groups experience both income poverty and 'time poverty' – defined as a lack of discretionary or meaningful free time after accounting for necessary paid work, unpaid work and personal care. Time poverty especially affects parents with children and notably single parents (Goodin et al. 2008). A study in the UK found 10 per cent of the population to be time-poor, 20 per cent income-poor and 1.6 per cent to be both time- and income-poor. However, this last group accounted for between 10 per cent and 14 per cent of all children in the UK and around one-half of single parents (Burchardt 2008). Ancillary social policies will evidently be needed to protect these groups.

This brings us to the role of the state. The state can in theory provide support enabling hours of work to be reduced faster than productivity growth (Pullinger 2013). But the influence of government extends wider than this. It defines expectations about who should seek paid work and the normal age of entry into and retirement from the labour force. Most of the activation and welfare-to-work policies of recent years have been about increasing the proportion of the population who are employed, in particular bringing in women (lone parents, but also partnered mothers) and people with disabilities. Reduced work can be achieved by both

---

[13] RWT is also likely to affect the rate of productivity growth. Most studies show that reducing hours of work improves hourly productivity. This means that the carbon savings from RWT will be somewhat less than shown in those studies assuming no such productivity effect.

[14] There is another option: household consumption can be maintained despite RWT if it is financed by borrowing. The tremendous rise in household debt since 1980 has provided a way to mitigate stagnant real wages by passing the burden forward to future generations. If this continues the climate benefit of RWT could be nullified. The association between debt and emissions has been mentioned in Chapter 3.

reduced working time and lower rates of participation. Issues of gender and age equity arise. Introducing RWT whilst maintaining the current retirement age and pension benefits would place the immediate burden on working-age households and children. Raising the average age of retirement plus gradually reducing the working week in older age would go some way to rebalancing this – and would break down the sharp division between work and retirement with other wellbeing benefits (Coote et al. 2010).

Finally, a case can be made that owners of capital should somehow contribute to and fund shorter hours of work. RWT plus no change in hourly rates of pay redistributes income within the paid workforce, but does nothing to redistribute from capital to labour (Kallis et al. 2013). This case is all the more important given the arguments above that lower growth will boost the share of capital in the economy and reduce the share of labour. The issues of justice and fairness resurface: redistribution from capital to labour and from high incomes to low will become a necessity in post-growth.

In conclusion, reducing the extent of time in paid work remains the surest strategy to move an economy towards a steady state. The major reason is simply that real incomes would grow less fast than they otherwise would. RWT provides a transitional programme that can be introduced gradually over time. Yet RWT directly contradicts dominant economic pressures to work longer hours and dominant social policies to *extend* paid work and to relate social benefits more closely to labour force participation. It will also require careful design to ensure that the lower-paid and time-poor are not disadvantaged.

CONCLUSION: POST-GROWTH AND SUSTAINABLE WELLBEING

This chapter has moved to the third level of decarbonisation. Going beyond the green economy and recomposing consumption it would decarbonise by reducing aggregate demand in rich countries. This post-growth scenario envisages a steady-state economy, the achievement of which will require a degree of negative growth or degrowth in the richest nations.

This implies a very different type of economy to today's: one where the emphasis is on reproduction not production, investment not consumption, more discretionary time not more commodities, more equality and redistribution not less. It will pose new questions for the welfare state and require new institutional ways of combining sustainable consumption with equity and justice. It will also require a new conception of the economy, one embedded in social relations, which are in turn embedded in the planetary environment.

But post-growth also poses a severe test for current social policies and the monetary aspect of wellbeing (which will continue to be important). It will put further pressures on the distribution of market incomes and the sustainability of social security programmes. The only solution that I can see for a 'post-growth welfare state' is to spread wealth more evenly through a variety of radical policies covering taxation and ownership. This leads to further speculation on minimum and maximum incomes, a revived core economy and a revitalised public sphere. An economy and a society that can no longer rely on annual growth will require a radical redistribution of carbon, time and the ownership of wealth (Coote and Franklin 2009).

The most realistic policy to achieve this transition, I argue, is gradually to reduce paid work time, and thus absolute levels of incomes, consumption and emissions. RWT constitutes another piece of the eco-social policy jigsaw, one that can improve sustainability as well as other dimensions of wellbeing. Table 8.2 summarises some of the novel eco-social policies needed to bring about this transition, as well as the transformations of existing social policy goals.

How likely is all this? The entire scenario must seem utopian. Yet is it not possible that we are entering a phase of post-growth by accident rather than design? Growth has slowed markedly since 2008, and some theories and models predict further slowdowns in the OECD (Gordon 2012). No renewable technologies can match the energy efficiency of carbon-based technologies in the heyday of cheap oil. Productivity is likely to decline further owing to the ongoing spread of the service economy, declining stocks of fossil fuel, materials, water and land, and the depredations of climate change itself.

Post-growth is a revolutionary idea that betokens the end of capitalism, at least in its original birthplaces in the North if not across the world. But perhaps capitalism is already running out of steam and slowly slumping towards a steady-state economy (Streeck 2016). If so, the prospect is decidedly bleak for human wellbeing. A sustainable post-growth economy that avoids dangerous climate change and enhances welfare will not come about by accident; it will require design.

Table 8.2 *Summary of eco-social policies in Chapter 8*

	Present idealised welfare goals	C1: Ramp up eco-efficiency	C2: Recompose consumption	C3: Post-growth: reduce consumption
S1: Redistribute income	Guarantee a minimum-income floor. Reduce inequality.	Compensate carbon-pricing losers. Manage inequality to enhance the effectiveness of carbon pricing.	Reduce inequality to avoid positional consumption and hyper-consumption races.	New prominence for redistribution. Minimum-income floor. Maximum-income ceiling.
S2: Social consumption	Publicly provide vital need satisfiers and discourage 'bads'.		Develop social consumption. Decarbonise welfare states.	New prominence for collective goods and services.
S3: Social investment	Develop human and social capabilities.	Build personal and social capabilities for green new deal programmes.	Extend citizen participation and control re energy and utilities.	Develop core capabilities and autonomy through and for the core economy.
ES: Novel eco-social policies		ES1: Green new deal: house retrofitting and allied programmes. ES2: Social tariffs for energy and water.	ES3: Regulate advertising, especially directed at children. ES4: Tax high-carbon luxuries: smart VAT, etc. ES5: Trial personal carbon rationing. ES6: Develop upstream prevention throughout public policy.	ES7: Reduce working time. ES8: Expand collective ownership of wealth and capital, starting with energy supply. ES9: Foster core economy and co-production. ES10: Develop population policy.

9. Conclusion: a three-stage transition

THE GLOBAL ENVIRONMENT

How can we achieve a fair distribution of wellbeing, while staying within environmental limits? Three critical biophysical boundaries are already overreached. One of these, global warming (more often called climate change), threatens catastrophic consequences for human society. Our overriding goal must therefore be to remain within what Kate Raworth has called a 'safe and just space for humanity': to improve human wellbeing across the globe whilst at the same time preserving the preconditions for a safe planet and the future wellbeing of its peoples. This is a huge challenge. The 2015 Paris climate agreement, though welcome, leaves a yawning gap between where we are now and where we need to go.

We must first establish a clear understanding of what constitutes human wellbeing and how it can be measured. Without this to guide policy and practice, there is a real danger that efforts to guard against catastrophic climate change will undermine wellbeing and widen inequalities. I argue in Chapter 2 that only a concept of universal human needs can provide such an understanding. It enables comparisons across cultures today and between the present and the future. In normative terms needs trump wants and so offer a guide to some difficult choices today, especially in affluent societies. Needs imply a standard of sufficiency, rather than maximisation. They imply reasonable equality above sufficiency thresholds, rather than indifference towards inequality. They make it possible to interrogate and modify existing social welfare goals, such as the UN Sustainable Development Goals, and to strengthen their logic, viability and appeal.

However, the pursuit of social, egalitarian and sustainable goals must contend with a political economy driven fundamentally by greed and the pursuit of profit. Indeed, capitalism is *the* systemic driver of climate change. Unless powerful countervailing forces are in play it also leads to unacceptable levels of inequality and opportunities for predatory practices by the powerful. Growing inequalities within nations since 1980 have exacerbated consumerism and GHG emissions. As Chapter 3 makes clear, both the ecological sphere and the social sphere are threatened by a runaway economic system of capital accumulation. Against this the technological dynamism of capitalism can continually improve labour productivity and, later and more slowly, the 'eco-efficiency of production' – the 'green growth' strategy. But it will need to do so at an unimaginable rate over the next three decades to have a good chance of staying within climatic boundaries.

Furthermore, eliminating poverty on a world scale can only be squared with planetary sustainability if the current model of economic growth is abandoned. If the business-as-usual model were used to eliminate poverty it would devastate the planet. All strategies to eliminate global poverty are untenable unless the poor get a bigger slice of the whole cake, but there are limits to its expansion because of global constraints on emissions.

This tragic global dilemma is best tackled by introducing need-based criteria. Using a range of non-monetary measures of need satisfaction, we calculate in Chapter 4 a different measure – the 'emissions efficiency of wellbeing' – across countries and time periods. The results are mixed but revealing: certain middle-income countries actually manage to combine a reasonable floor level of wellbeing with sustainable emissions. Other countries are heading for a fast leap in emissions if they are to attain reasonable standards of need satisfaction. What is more, the Paris agreement is founded on much wishful thinking about climate mitigation, including unproven technologies, such as bioenergy and carbon capture and storage, with uncertain social consequences.

Three conclusions follow. First, *consumption* and consumption-based emissions, ignored by the green growth agenda, must be given equal priority, especially in the rich world. Second, issues of global equity, almost entirely absent from international climate negotiations so far, must at the very least be discussed and confronted. Ideally a global agreement on fair burden-sharing based on contract-and-converge principles should be negotiated. It will certainly confront profound political and systemic obstacles, but sooner rather than later it must be taken seriously if climate justice is to be pursued. Third, 'affluence' has a class as well as a national dimension: there are also wealthy high-emitting classes in poor countries. Some way needs to be found to combine these concerns of domestic justice with both international and intergenerational justice in a global equity framework. A robust concept of human need will play an essential part in resolving these dilemmas.

Green growth forms the centrepiece of the Paris agreement and is the only current politically viable strategy for a global low-carbon economy. There are several critical reasons for believing that green growth alone cannot meet the agreed decarbonisation targets. But it also suffers from an inherent social dilemma: it has to operate within a global economy of gross inequality (both between and within countries). As a consequence, it will in practice fail to raise global standards of wellbeing to a sufficient level, or will fail to reduce emissions at a sufficient rate, or will fail at both.

TOWARDS ECO-SOCIAL POLICIES IN THE AFFLUENT WORLD

The rich world bears a double obligation: it has been responsible for the majority of accumulated greenhouse gases to date; and because of its greater wealth it has far greater capacity to lead the world in fast decarbonisation. We must also recognise that the countries of the rich world are themselves riven by inequalities in wealth, income and political power.

Over the second half of the twentieth century, citizens (and some non-citizens) of rich countries have accumulated means of satisfying their needs, including health, education and a modicum of income security through the development of national welfare states. Provision according to need has always been in tension with capitalist values and private sector interests, and has had to be constantly fought for and renewed: an evolving war of attrition heightened by the financial crisis of 2008 and its aftermath. Yet, at the same time, social policy has helped to maintain conditions for the reproduction and survival of the market system as a whole – a role that has so far prevented its dissolution.

I argue in Chapter 5 that there are three implicit goals of welfare states – redistribution, social consumption and social investment – that can provide a framework for understanding their future roles in the age of the Anthropocene. This framework is then applied in the remainder of Part II of the book.

Welfare states are essentially national creations. Climate change now imposes on them new threats but above all the urgent need to decarbonise their economies by around the middle of this century. Almost every nation state in the world has now produced an intended nationally determined contribution (INDC) stating its goals to reduce emissions. However tardy and inadequate they are, these INDCs indicate that, as well as moral pressures to act collectively to save the planet, there are internal economic and political reasons why states will aim to modify markets to pursue climate goals. This is the logic underpinning today's dominant strategy of 'green growth'.

But, just as with state commitments to welfare, there exist tensions between evolving 'climate states' and their existence as guardians of national competitiveness within a profit-driven capitalist world economy. Comparative research has revealed persistently different national patterns in both domains. Distinct 'welfare regimes' exist between the Nordic countries, the founder members of the EU and the English-speaking countries. And these differences are roughly mirrored by variations in the intensity and achievements of climate mitigation programmes.

In Chapter 6, I take for granted the arguments for green growth and mainstream climate mitigation policies and investigate their social consequences. First, raising the cost of using carbon, whether directly, as in domestic heating and transport, or indirectly, as in more general carbon taxes or carbon caps, will almost always have a regressive impact – reducing the incomes of poorer households more than those of richer households. Second, compensating such 'losers', the traditional answer advocated by economic theory, is difficult to do, incurs economic costs, and absorbs carbon revenues that could be spent more productively elsewhere. Yet, third, to redistribute incomes downwards in a more general way, a traditional goal of social policy, may increase aggregate emissions if unaccompanied by other measures. The usual combination of pricing plus state transfers is not a way forward, at least not in isolation.

One conclusion is that welfare states in the Anthropocene will need to forge new policy instruments. These are 'eco-social policies', defined as policies that simultaneously and explicitly pursue both equity/justice and sustainability/sufficiency goals. I set out several examples, such as targeted retrofitting of dwellings and social tariffs for domestic energy. A second conclusion is that successfully implementing eco-social policies would require a more dirigiste, state-led strategy: an active interventionist 'innovation state', with substantial public investment, state banking, subsidies and other incentives to private investment and greater regulation and planning. It would also require a variety of social and economic measures to offset regressive impacts on lower-income groups and to reverse growing levels of inequality. The investment functions of social policy would be enlarged and more closely integrated with environmental investment. This would require tipping the balance of the economy from private to state investment.

Green growth focuses on the eco-efficiency of production and pays no attention to the essential links between patterns of *consumption* and GHG emissions. A substantial proportion of greenhouse gases associated with goods consumed in the rich world are actually emitted in other countries, where the goods are made. This means they are off the balance sheet for affluent consumer countries, as though they were not responsible for them. With this in mind, I turn next, in Chapter 7, to consumption-based emissions in affluent countries and explore routes towards lower-carbon consumption. I propose a new meta-goal for policy: to 'recompose' consumption in affluent societies in order to develop a safe 'consumption corridor' between *minimum* standards, allowing every individual to live a good life, and *maximum* standards, ensuring a limit on every individual's use of natural and social resources.

To achieve this goal of fair sustainability we need a method to distinguishing 'necessary' consumption from 'luxury' and 'locked-in' consumption and ways of living. Returning to human need theory I propound a novel methodology – the dual strategy – for identifying need 'satisfiers'. This brings together citizens and experts in groups to identify goods and services necessary for a flourishing and social life in particular social contexts. Such a procedure is now increasingly used to define a 'decent minimum' standard of living across the European Union. Ways of extending this method to define 'maximum' standards are then discussed. In this way a collective means of challenging established patterns of consumption might be nourished.

In the light of powerful corporate and other interests shaping consumer preferences Chapter 7 also advocates a broad strategy of upstream prevention: to move health and other social interventions upstream to prevent social maladies emerging in the first place. This points the way to integrate preventive climate and social programmes, and further eco-social policies are suggested: regulating advertising, taxing high-carbon luxuries, rationing carbon at the household level, and socialising some high-carbon services. All would confront currently dominant ideas of consumer and market sovereignty. They would require social mobilisation plus governments willing to pursue an upstream and integrated preventative strategy embracing economic, social and environmental goals. To recompose consumption in this way will require new forms of 'eco-welfare state' at the national level.

Recomposing consumption in this way would take us a step further towards sustainable wellbeing, beyond green growth. But even taken together I believe the strategies could not reduce emissions in rich countries far or fast enough to achieve a safe climate. I therefore turn in Chapter 8 to *post-growth* as the only remaining option. It envisages a post-growth steady-state economy, the achievement of which will require a prior degree of negative growth or degrowth in the richest nations. The most realistic policy advanced to achieve this transition is gradually to reduce paid work time, and thus absolute levels of incomes, consumption and emissions. RWT constitutes another piece of the eco-social policy armoury, one that can improve sustainability as well as other dimensions of non-monetary wellbeing and human flourishing.

But degrowth threatens almost the entire suite of current social policies that support wellbeing in a market economy. All welfare states and social security programmes have been built on, and assume the continuance of, conventional economic growth in GDP. If this is removed their fiscal stability and political legitimacy would be threatened. At the same time lower growth would likely drive up further inequality in wealth and income. In this scenario *all* varieties of capitalism and *all* welfare regimes are unsustainable. All have been historically based on a high-energy, high-growth economic pathway (Koch and Fritz 2014).

The picture that emerges of a sustainable post-growth welfare system combines top-down and bottom-up mobilisations and action. To spread wealth more evenly through society, a range of radical top-down policies would be needed covering taxation and the ownership of wealth. New forms of public ownership and control would in any case be needed, including socio-natural resources such as energy and water. This could involve some combination of state and common ownership. The idea of the natural commons can be extended to that of a 'social commons' in order to protect and extend the inherited suite of social institutions that help maintain the social fabric of modern societies.

Much of this will need to be built up from below by resourceful communities. Preventing harm is arguably most effective when it involves change from the bottom up, with people and organisations becoming more proactive: building up their own immune systems, both literally and metaphorically, so that they become less susceptible to harm; and changing attitudes and capabilities so that they are better able to withstand harm by taking positive actions themselves (Coote 2012a). The unpaid core economy, so important in meeting basic needs and enabling wellbeing to flourish, must now be comprehensively recognised and supported both in economic models and in actual practice.

Taken together, Chapters 6, 7 and 8 amount to a three-stage transitional strategy for sustainable wellbeing: green growth, recomposed consumption, post-growth. The current stage of green growth, driven by rapid decarbonisation and improvements in eco-efficiency of production, is of major importance and drives the post-Paris agenda. It is aligned with core state imperatives in the contemporary world, economic growth and energy security among them. But it cannot be the end goal, since it will be environmentally unworkable and unjust.

Yet, at the same time, post-growth appears to be a political non-starter. 'An industrial system with reduced material demand is not in any group's direct interest, although it is essential to human survival' (Allwood et al. 2017). Too many powerful constituencies, not to speak of consumers and citizens, would be faced with material losses. Dominant interests, institutions and ideologies would be so threatened that the idea appears politically unfeasible. It is for this reason, among others, that I propose an interim strategy to recompose consumption in rich countries towards low-carbon necessities. It would begin to confront hyper-consumption and the ideology of consumer sovereignty. It would provide a route from the impossible present to a possible future.

Table 9.1 reprises Table 8.2. It relates these arguments and propositions to social policy, setting out a transitional three-stage strategy to move away from the present system of threatened traditional welfare states within unequal and unsustainable neoliberal forms of capitalism, towards more sustainable, equitable and welfare-enhancing alternatives. It summarises arguments built up through the previous three chapters, including both the new roles for traditional social policies and the suite of proposed new eco-social policies. The final two rows of Table 9.1, added here, begin to relate this schema to the global capitalist system.

THE POLITICAL ECONOMY OF THE THREE TRANSITIONS

Throughout this book I have sought to situate these eco-social changes in the context of the global political economy. This begs the question of whether the transition route to sustainable wellbeing that I have charted is achievable at all. Everything depends on the nature, variability, flexibility and reformability of capitalism.

Perspectives on this question vary widely. Some see capitalism per se as the root cause of the stupendous growth of commodity production, consumption, and despoliation of the planet. Examples include Naomi Klein's (2015) book subtitled *Capitalism vs. the Climate* and a corpus of 'eco-Marxist' works, such as O'Connor (1997), Burkett (1999) and Foster (2009). One proponent, Jason Moore, has challenged the very idea of the Anthropocene, claiming instead that we should speak of the *Capitalocene* – 'an ugly word for an ugly system' (Moore 2015).

Others focus on 'varieties of capitalism' that have emerged historically according to specific class interests, institutions and ideas. For example, Hall and Soskice (2001) distinguish between liberal market economies and coordinated market economies. This opens the possibility of some varieties of capitalism being more capable than others of accommodating or nurturing eco-social transitions to avert climate catastrophe.

C1: Coordinated Capitalism and Green Growth

Chapter 5 established that there are significant differences across the nations of the OECD in their welfare and climate regimes. The question then arises: do these connect to differences in their forms of capitalism? Hall and Soskice (2001) argue that both types of capitalist economy they

Table 9.1 *Summary of social policy, eco-social policies and forms of economy in the three stages of decarbonisation*

	Present idealised welfare goals	C1: Ramp up eco-efficiency	C2: Recompose consumption	C3: Post-growth: reduce consumption
S1: Redistribute income	Guarantee a minimum-income floor. Reduce inequality.	Compensate carbon-pricing losers. Manage inequality to enhance the effectiveness of carbon pricing.	Reduce inequality to avoid positional consumption and hyper-consumption races.	New prominence for redistribution. Minimum-income floor. Maximum-income ceiling.
S2: Social consumption	Publicly provide vital need satisfiers and discourage 'bads'.		Develop social consumption. Decarbonise welfare states.	New prominence for collective goods and services.
S3: Social investment	Develop human and social capabilities.	Build personal and social capabilities for green new deal programmes.	Extend citizen participation and control re energy and utilities.	Develop core capabilities and autonomy through and for the core economy.
ES: Novel eco-social policies		ES1: Green new deal: house retrofitting and allied programmes. ES2: Social tariffs for energy and water.	ES3: Regulate advertising, especially directed at children. ES4: Tax high-carbon luxuries: smart VAT, etc. ES5: Trial personal carbon rationing. ES6: Develop upstream prevention throughout public policy.	ES7: Reduce working time. ES8: Expand collective ownership of wealth and capital, starting with energy supply. ES9: Foster core economy and co-production. ES10: Develop population policy.
Welfare system Enabling political economy		*Productive welfare state Coordinated climate capitalism*	*Eco-welfare state Reflexive capitalism?*	*Sustainable wellbeing system Beyond capitalism: stationary-state economy*

identify can act to achieve national competitive advantage. In *coordinated market economies (CMEs)*, firms depend more heavily on non-market collaborative relationships to coordinate their endeavours with other actors and to construct their core competencies. In *liberal market economies (LMEs)*, firms coordinate their activities primarily via competitive market arrangements and formal contracting.

These differences are reflected in the organisation of labour markets, systems of skill formation and social policies. For example, CMEs are characterised by more generous wage replacement rates and strong employment protection, because these foster a high-skills production system requiring firm loyalty and lower rates of labour mobility. An extensive welfare state can level the ground between corporations and support more state–business coordination. On the other hand, the lower wage and more flexible labour markets characteristic of LMEs will benefit from, and tend to reproduce, lower and less rights-based welfare benefits. Similarly, equity-based pension systems, typical of liberal welfare regimes, arise from and reinforce liberal shareholder capitalism (Schröder 2009).

These relationships are illustrated in Figure 9.1. The liberal economies and liberal welfare regimes are found in the Anglosphere; coordinated economies are found in Europe and Japan and map on to distinct conservative-corporatist and social democratic welfare regimes. How far such correspondence is due to *institutional complementarities* and how far

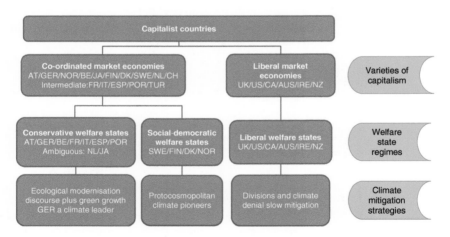

Source: Schröder (2009, fig. 2; last line added by IG).

Figure 9.1   Typologies of capitalism, welfare states and climate mitigation strategies

a result of historical path-dependency is the subject of debate (Streeck 2010).

Are these varieties of capitalism and welfare regime reflected in differing climate strategies? An active social policy, typical of CMEs, can contribute to successful climate action in several ways. First, it can provide a climate 'safety net', by socialising the risks of climate change and ensuring policy responses do not aggravate existing inequities or create new ones. Second, an extensive social investment programme in education and community capacities can contribute directly to the mobilisation of societal resources to confront climate change. Third, collectively provided welfare provides a more solid floor, enabling citizens to cope with turbulence and rapid change in a creative way. Without some floor of security, new risks tend to exacerbate social vulnerabilities while citizens are trapped and disempowered (Room 2016). Coordinated market economies associated with social democratic and some conservative or corporatist welfare states are more likely to regard economic, social and ecological values as mutually reinforcing (Gough et al. 2008a). They can achieve a more effective eco-social synergy without harming national competitiveness. Germany and the Nordic countries exemplify this approach.

For CMEs, the discourse of 'ecological modernisation', similar to green growth, provides an economic rationality for moving towards environmental sustainability (Christoff 1996; Mol et al. 2009). It facilitates policy 'concertation' or integration across environmental, economic and social issues (Nilsson and Eckerberg 2007). Going further, a more active state, with public investment, public banks, and greater public support for industrial policies and research and development, can pursue a stronger form of ecological modernisation (Jacobs and Mazzucato 2016) and ideally can integrate policy both horizontally and vertically. Horizontal integration would connect previously disparate sectors, ministries and agencies, such as energy, social protection and finance. Vertical integration would devolve power and utilise the local knowledge of communities and local government while coordinating these at the national and supranational level (Lafferty and Hovden 2003).

By contrast, the more liberal, market-oriented economies of the English-speaking world combine weaker social policy effort with significant levels of climate denial and less consistent attention to climate mitigation (Dryzek et al. 2003; Meadowcroft 2005; Gough et al. 2008b; Christoff and Eckersley 2011). This has been recently illustrated by US President Trump's appointment of outright climate deniers to cabinet posts.

Of course, these patterns are not set in stone, and the future is unpredictable. If the Eurozone fails and the EU weakens dramatically, the best present hopes we have for an integrated eco-welfare state will be dashed.

Nevertheless, it seems highly unlikely that the liberal, business-as-usual (BAU) scenario that is reckless of climate change can succeed as a credible medium-term strategy. Defending an out-dated carbon economy will eventually threaten core state imperatives such as economic growth and energy security. Green business interests will challenge BAU, gaining first-mover advantage by forging a renewables-based economy. Above all, BAU would speed up destruction of planetary systems, threatening aspects of human wellbeing and social functioning on which profits ultimately depend.

It is true that none of the rich economies, whatever the welfare regime or form of capitalism, comes close to sustainable levels of consumption-based emissions (Koch and Fritz 2014). There is nevertheless considerable argument and evidence that eco-efficient green growth policies (C1) can be rolled out together with social equity goals more effectively in states where there are coordinated rather than liberal varieties of capitalism.

C2: Reflexive Capitalism?

Moving to the next stage of the transition, is it possible to envisage a form of capitalism that enables governments and citizens to shift consumption patterns in a sustainable direction? Unlike the green growth scenario, this stage challenges current ideas of prosperity and consumer sovereignty by advocating cuts in high-carbon luxury consumption. It involves scrutinising Western lifestyles and individualistic consumer culture, though not yet the commitment to economic growth.

If a capitalist economy is to evolve that is capable of recomposing consumption (C2) it will arguably need to develop three characteristics: reflexivity, a commitment to prevention, and a capacity to integrate local (bottom-up) and national (top-down) agency.

Reflexivity

Reflexivity signals the ability of a structure, process or set of ideas to change itself in response to reflection on its performance. In institutional analysis, it is a potential antidote to the problematic path-dependency in which we find ourselves (Beck et al. 1994). Some argue that a global environmental crisis requires a set of social institutions that recognise the Earth system as a key player, and that respond not just to present human voices but also to future generations and other components of social-ecological systems that have no voice. Dryzek (2016) claims that this form of *ecosystemic reflexivity* is the first virtue of political institutions in the Anthropocene. Furthermore, reflexivity is required to prioritise needs over preferences. Since our wants are shaped by structures, interest groups and private and public institutions, only a reflexive discourse can help us escape the appeal to consumer sovereignty.

To build a reflexive political culture, new methods are needed to embed deliberative dialogue in decision making. In Chapter 7 I advocate a dual strategy, which brings together lay people and experts, combining codified and experiential knowledge to consider how to cut emissions in ways that are both fair and effective. There is some evidence that this approach can enlarge the perspectives of all participants and build consensus around shared interests as well as shared responsibilities towards future generations and planetary resources.

Prevention
Reflexivity is one of the preconditions for a *precautionary* and *preventive* approach, which has featured throughout this book. The need to establish a moral and technical foundation for preventive policy has gained new urgency from the environmental crisis: it is the ultimate 'serious harm' that threatens human wellbeing. As noted in Chapter 7, prevention requires a reflexive understanding of social causation in complex social contexts, and an ability to look ahead, exert critical autonomy, criticise taken-for-granted consumption practices and devise alternatives.

There are two further preconditions for effective prevention – a scientific understanding of cause and effect, and a capacity for controlled public intervention in social life. Current socio-economic institutions are not best designed to meet these conditions. The obstacles are formidable. They are located in the economy, sustained by neoliberal ideology, and reinforced by the private interests and institutions that benefit from them.

Keynesianism, which dominated until the mid-1970s, founded what could be regarded as a precautionary, preventive style of economic management. However, anomalies within this system began to accumulate, which resulted in it being displaced, initially in the English-speaking world, with what is commonly referred to as neoliberalism (Hall 1993). It is worth noting that there are two intellectual 'counter revolutions' here: Hayekism, which holds that society or its representatives cannot conceive of causal connections between policies and outcomes in a scientific way; and neoclassical economics, which has abandoned the idea of a distinct macro-economic method and advanced the belief that markets are powerfully self-stabilising.

Shaped by these ideas, the current neoliberal economic model hinders the necessary public regulatory, fiscal and mobilising initiatives that are essential for a coordinated social–environmental–economic preventive strategy. And this infects other policy domains. Within social policy, for example, 'contracting out' spawns a growing network of private providers and new forms of private-interest government. Within climate change mitigation, it has seen off effective carbon taxes and new public investment in ecosystem maintenance (Gough 2015b).

Following the 2008 crisis, the case for a new sustainable economic model has gained some ground. From this it follows that government should pursue preventive economic policies, including substantial restructuring of financial markets, active macro-economic management and some socialisation of investment. More economists now recognise the dangers of business as usual, and a few have gone further, making explicit distinctions between good and harmful (or 'socially useless') economic activity (Turner 2012). Nevertheless, the solid reality remains, in the words of Colin Crouch (2011), 'the strange non-death of neoliberalism'.

Integrating local and national agency
While neoliberal economics severely limits the scope of national top-down agency, it may leave some kinds of bottom-up agency relatively undisturbed. Much green political thought has tended to view states as part of the problem rather than the solution. It applauds participation and community-led action as the way towards sustainable policy, advocating both local and communal forms of organisation and regulation, and the need for global networks of citizen action (Barry 1999).

Pressures from below in the form of locally generated campaigns have been significant drivers of welfare state expansion and have helped to build momentum for environmental change. Whether these can be replicated on a global scale to pressure nation states and global institutions remains doubtful. Klein (2015) reviews historical examples, noting the victories of the US civil rights movement and the global women's movement. But she concludes that 'sharing legal status is one thing; sharing resources quite another'.

While local campaigns usually hope to achieve change by influencing regional and national governments, most local prefigurative initiatives aim to show what can be done without state interference. The dominant motive is to improve wellbeing on the ground, via such policies as controlling air pollution, improving housing, and providing better public transport, parks and green spaces. In so far as effects are scaled up, this is expected to happen not through national or trans-national agency, but through proliferation and independent coordination of local initiatives.

The transition towns movement explicitly bills itself as a response to the failure of higher levels of government to confront resource constraints and climate change. Networks across localities – such as the International Council for Local Environmental Initiatives cities network – also provide sites for deliberation (Bulkeley 2013). Ostrom (2009) gives many examples of 'non-global' initiatives contributing to climate mitigation, from US states to Canadian cities. Mayne (2017) shows how these can be adapted and adopted to work in low-income areas in the UK.

Yet there are strong arguments that local campaigns and prefigurative action cannot be *the* approach to climate mitigation and that stronger top-down state initiatives are essential. In Scandinavia, not only were many local initiatives kick-started by central programmes and funding, but research has shown that when these ran out the local initiatives demonstrated relatively little staying power. Lafferty (2012) argues that, when faced with the overwhelming and imminent threat of climate change, centralised authoritative responses backed by constitutional guarantees are essential. Rules, regulation, precautionary monitoring and more should be part of the central state arsenal.

The challenge, then, is to achieve more vertical integration, through dynamic interaction between local and national agency. This would involve building trust and generating far greater capacity to engage and work together, so that top-down and bottom-up transformations are mutually reinforcing rather than separate (Scoones et al. 2015). A recursive or experimentalist architecture is called for whereby local initiatives are encouraged but these are then compared and evaluated in peer reviews. At the same time national framework goals, measures and procedures are themselves periodically reviewed and revised in an iterative process (Sabel and Zeitlin 2008).

There are pessimistic views about the capacity of capitalist democracies to take effective action on climate change, let alone to support such levels of reflexivity, not only from neo-Marxists. Dahl (1998) speaks of the relationship between a country's democratic political system and its non-democratic economic system as a formidable and persistent challenge. For Lafferty (2012) competitive democracy is a major roadblock confronting moves to a sustainable future. These positions now gain added force from the closer fusion of plutocracy and the state. In the current climate eco-localism can mean more privileged communities disengaging from wider struggles, whereas what is needed is for pressures from below to engage with and inform the reform of the wider political economy in the interests of wider equality and social justice (Albo 2007). One example here is Colin Hines's (2017) case for 'progressive protectionism' to enable a re-localisation of economic life.

Yet Eckersley (2016) qualifies this negative view, finding evidence in Norway and Germany of a more reflexive, cosmopolitan discourse and a governmental recognition of our common global dilemmas and the need for cooperative global action. If she is correct, there are glimpses of hope that some capitalist states could be capable of shifting consumption norms. If capitalism survives because it continually adapts to changing circumstances, then it is possible to imagine a move in this direction that does not threaten the core interests of capital and states in accumulation and growth. There are several qualifying conditions. First, the political economy would be that of coordinated market capitalism. Second, it would have embraced green growth as a positive experience. Third, it would evolve by developing reflexivity, a commitment to prevention and the capacity to integrate local and national agency.

This would require a move beyond existing forms of coordinated capitalism. Just as capitalist economies embraced redistributive social policy and the welfare state in the second half of the twentieth century, so the *composition of consumption* would need to enter political debate and policy output in the twenty-first century. This is speculation. So, too, is the next stage of transition.

C3: Post-Growth

I have argued that since neither green growth (C1) nor recomposed consumption (C2) will reduce emissions far enough or fast enough to avert catastrophic climate change, the only remaining option is to move beyond growth to a steady-state economy (C3). Does this imply the end of capitalism? Marxists clearly answer yes, but so do others. Jeremy Grantham, global investment manager and funder of the Grantham Institute at the LSE, says: 'Capitalism, by ignoring the finite nature of resources and by neglecting the long-term well-being of the planet and its potentially crucial biodiversity, threatens our existence' (in Klein 2015: 233).

As an *economic system*, capitalism is differentiated from others by its drive to accumulate and its inbuilt tendency to expand. As such it is widely seen to be on a collision course with the Earth system. 'There is no way out of the dilemma between the capitalist imperative to accumulate and the limits that nature sets' (Altvater, cited in Demaria et al. 2013). Moving beyond growth challenges capitalism in (at least) two ways. First, it directly confronts the innate drive to accumulate capital. Second, to avoid a devastating impact on material living standards it will be necessary to question and partially dismantle a defining feature of capitalism: the private ownership of the means of production.

Some post-growth economists, such as Herman Daly and Tim Jackson, are more ambivalent. They consistently stress that profound transformations are needed in the current growth-obsessed economic system, but demur as to whether capitalism can survive at all. 'The economy of tomorrow calls on us to revisit and reframe the concepts of productivity, profitability, asset ownership and control over the distribution of surpluses ... Perhaps we could agree to coin the term "post-growth capitalism". Perhaps not . . .' (Jackson 2017; cf. Jackson and Victor 2015).

In my view post-growth is incompatible with any form of capitalism – or at least with capitalism as a global system. If it happens, the process of moving beyond growth could possibly begin in the rich world: these privileged zones might continue to trade with the rest of the world whilst developing a steady-state economy. But ultimately there would likely need to be a move away from global economic integration by free trade, free capital mobility and export-led growth – and toward a more nationalist orientation that seeks to develop domestic production for internal markets as the first option (Daly 1996; Hines 2017).

However, the future may not work out this way: the sequence of transitional stages that I have envisaged may not come about. Failure of climate policy at the global level could see temperatures rising by 3 °C before the end of this century. Then global warming would threaten the core productive capabilities and overwhelm basic institutional preconditions for basic needs and human security. The geographic spread of these effects would place quite different demands on countries according to their spatial location and their prior institutional structures (Buch-Hansen 2014). But it would certainly transform the nature of political economies and state functions across the globe.

Writers such as John Urry (2011) and Peter Christoff (2013) imagine that the failure of transition would bring about *fortress states*. These would oversee but could not effectively manage survival in a permanent state of emergency. The priority goals would be adaptation to a hotter and more unstable climate, entailing policies to secure and maintain supplies of energy, water and food. The dominant political narrative would be survival and minimal national welfare. The maintenance of borders and social order would require new policing powers. There would be little scope for any kind of welfare state let alone social investment. For the vast majority, opportunities for human flourishing would decline.

This book has been unashamedly premised on an older assumption of optimism and human progress, but we should be aware of these darker alternatives. Hopefully this nightmare vision will spur us to work harder to achieve the more optimistic transitions to a needs-based economy and society set out in this book.

References

Advani, A., Johnson, P., Leicester, A. and Stoye, G. 2013. *Household Energy Use in Britain: A Distributional Analysis*. London: Institute for Fiscal Studies.

Albo, G. 2007. The limits of eco-localism. *Socialist Register* 2007, 43, 337–363.

Alkire, S. 2002. *Valuing Freedoms: Sen's Capability Approach and Poverty Reduction*. Oxford University Press, Oxford.

Allwood, J.M., Ashby, M.F., Gutowski, T.G. and Worrell, E. 2013. Material efficiency: Providing material services with less material production. *Philosophical Transactions of the Royal Society*, 371, 20120496.

Allwood, J.M., Gutowski, T., Serrenho, A., Skelton, A. and Worrell, E. 2017. Industry 1.61803: The transition to an industry with reduced material demand fit for a low carbon future. *Philosophical Transactions of the Royal Society A: Mathematical, Physical and Engineering Sciences*, 375.

Anand, S. and Sen, A. 2000. Human development and economic sustainability. *World Development*, 28, 2029–2049.

Andersen, M.S. and Ekins, P. (eds). 2009. *Carbon-Energy Taxation: Lessons from Europe*. Oxford University Press, Oxford.

Anderson, K. 2012. Climate change going beyond dangerous – brutal numbers and tenuous hope. *Development Dialogue*, 61, 16–39.

Anderson, K. and Peters, G. 2016. The trouble with negative emissions. *Science* 354(6309), 14 Oct 2016, 182–183.

Athanasiou, T., Kartha, S. and Baer, P. 2014. *National Fair Shares: The Mitigation Gap – Domestic Action and International Support*. EcoEquity, Berkeley, CA and Stockholm Environment Institute, Stockholm.

Atkinson, A.B. 2015. *Inequality*. Harvard University Press, Cambridge, MA.

Bacon, R.W. and Eltis, W. 1976. *Britain's Economic Problem: Too Few Producers*. Macmillan, London.

Baer, P. 2013. The Greenhouse Development Rights framework for global burden sharing: Reflection on principles and prospects. *Wiley Interdisciplinary Review of Climate Change*, 4, 61–71.

Baiocchi, G., Minx, J. and Hubacek, K. 2010. The impact of social factors and consumer behavior on carbon dioxide emissions in the United Kingdom. *Journal of Industrial Ecology*, 14, 50–72.

Barbier, E.B. 2010. *A Global Green New Deal: Rethinking the Economic Recovery*. Cambridge University Press, Cambridge.

Barry, J. 1999. *Rethinking Green Politics: Nature, Virtue and Progress.* Sage, London.

Barry, J. 2002. Vulnerability and virtue: Democracy, dependency, and ecological stewardship. In B.A. Minteer and B.P. Taylor (eds), *Democracy and the Claims of Nature: Critical Perspectives for a New Century*, pp. 133–152. Rowman & Littlefield, Lanham, MD.

Barry, J. 2012. *The Politics of Actually Existing Unsustainability: Human Flourishing in a Climate-Changed, Carbon-Constrained World*. Oxford University Press, Oxford.

Barry, J. and Eckersley, R. 2005. W(h)ither the green state? In J. Barry and R. Eckersley (eds), *The State and the Global Ecological Crisis*, pp. 255–272. MIT Press, Cambridge, MA.

Bassi, S. and Averchenkova, A. 2016. *Beyond the Targets: Assessing the Political Credibility of Pledges for the Paris Agreement*. Grantham Research Institute on Climate Change and the Environment, London.

Bassi, S., Fankhauser, S., Green, F. and Nachmany, M. 2014. *Walking Alone? How the UK's Carbon Targets Compare with Its Competitors'*. Policy Paper. Centre for Climate Change Economics and Policy/Grantham Research Institute on Climate Change and the Environment, London.

Baumol, W.J. 1990. Entrepreneurship: Productive, unproductive, and destructive. *Journal of Political Economy*, 98, 893–921.

Baxter, J.L. and Moosa, I.A. 1996. The consumption function: A basic needs hypothesis. *Journal of Economic Behavior and Organization*, 31, 85–100.

Beck, U., Giddens, A. and Lash, S. 1994. *Reflexive Modernization: Politics, Tradition and Aesthetics in the Modern Social Order*. Stanford University Press, Stanford, CA.

Béland, D. and Mahon, R. 2016. *Advanced Introduction to Social Policy*. Edward Elgar Publishing, Cheltenham, UK and Northampton, MA, USA.

Bernays, E.L. 1955. *The Engineering of Consent*. University of Oklahoma Press, Norman.

Berners-Lee, M. and Clark, D. 2013. *The Burning Question*. Profile Books, London.

Blackburn, R. 2011. Reclaiming human rights. *New Left Review*, 69, 126–138.

Blackburn, R. 2013. *The American Crucible: Slavery, Emancipation and Human Rights*. Verso Books, New York.

Boardman, B. 2010. *Fixing Fuel Poverty*. Earthscan, London.

Boarini, R., Kolev, A. and McGregor, A. 2014. *Measuring Well-Being and Progress in Countries at Different Stages of Development*. OECD Development Centre Working Papers No. 325. OECD, Paris.

Bocchiola, M. 2013. Milanovic on global inequality and poverty. *Global Policy*, 4, 209–210.

Bollier, D. 2016. *Commoning as a Transformative Social Paradigm*. Available at: http://thenextsystem.org/wp-content/uploads/2016/04/DavidBollier. pdf.

Boulding, K. 1966. The economics of the coming Spaceship Earth. In H.E. Daly (ed.), *Environmental Quality Issues in a Growing Economy*. Johns Hopkins University Press, Baltimore, MD.

Bowen, A. and Hepburn, C. 2014. Green growth: An assessment. *Oxford Review of Economic Policy*, 30, 407–422.

Bowles, S. 2012. *The New Economics of Inequality and Redistribution*. Cambridge University Press, Cambridge.

Bowles, S. and Park, Y. 2005. Emulation, inequality, and work hours: Was Thorsten Veblen right? *Economic Journal*, 115, F397–F412.

Bows, A. and Barrett, J. 2010. Cumulative emission scenarios using a consumption-based approach: A glimmer of hope? *Carbon Management*, 1, 161–175.

Boyce, J. 2007. Inequality and environmental protection. In J.-M. Baland, P. Bardhan and S. Bowles (eds), *Inequality, Cooperation and Environmental Sustainability*, pp. 314–348. Princeton University Press, Princeton, NJ.

Boyd, Rodney, Turner, Joe and Ward, Bob. 2015. *Intended Nationally Determined Contributions: What Are the Implications for Greenhouse Gas Emissions in 2030?* Grantham Research Institute on Climate Change and the Environment, London.

Bradshaw, J., Middleton, S., Davis, A., Oldfield, N., Smith, N., Cusworth, L. and Williams, J. 2008. *A Minimum Income Standard for Britain: What People Think*. Joseph Rowntree Foundation, York.

Brandstedt, E. 2013. *The Construction of a Sustainable Development in Times of Climate Change*. Lund University, Lund.

Braybrooke, D. 1987. *Meeting Needs*. Princeton University Press, Princeton, NJ.

Brock, G. 2009. *Global Justice: A Cosmopolitan Account*. Oxford University Press, Oxford.

Buch-Hansen, H. 2014. Capitalist diversity and de-growth trajectories to steady-state economies. *Ecological Economics*, 106, 167–173.

Büchs, M., Bardsley, N. and Duwe, S. 2011. Who bears the brunt? Distributional effects of climate change mitigation policies. *Critical Social Policy*, 31, 285–307.

Bulkeley, H. 2013. *Cities and Climate Change*. Routledge, Abingdon.

Burchardt, T. 2008. *Time and Income Poverty*. CASE Report No. 57. Centre for the Analysis of Social Exclusion, London School of Economics, London.

Burkett, P. 1999. *Marx and Nature: A Red and Green Perspective*. St. Martin's Press, New York.

Cahn, E.S. 2000. *No More Throw-Away People: The Co-Production Imperative*. Essential Books, London.

Campiglio, E. 2015. *Beyond Carbon Pricing: The Role of Banking and Monetary Policy in Financing the Transition to a Low-Carbon Economy*. Grantham Research Institute on Climate Change and the Environment, London.

Caney, S. 2012. Just emissions. *Philosophy and Public Affairs*, 40, 255–300.

Caporaso, J.A. and Levine, D.P. 1992. *Theories of Political Economy*. Cambridge University Press, Cambridge.

Castles, F.G., Leibfried, S., Lewis, J., Obinger, H. and Pierson, C. (eds). 2010. *The Oxford Handbook of the Welfare State*. Oxford University Press, New York.

Chakravarty, S. and Tavoni, M. 2013. Energy poverty alleviation and climate change mitigation: Is there a trade off? *Energy Economics*, 40, S67–S73.

Chakravarty, S., Chikkatur, A., de Coninck, H., Pacala, S., Socolow, R. and Tavoni, M. 2009. Sharing global CO_2 emission reductions among one billion high emitters. *PNAS*, 106, 11884–11888.

Chancel, L. and Piketty, T. 2015. *Carbon and Inequality: From Kyoto to Paris*. VOX: CEPR's Policy Portal, Paris.

Chibber, V. 2013. *Postcolonial Theory and the Specter of Capital*, 1st edn. Verso Books, London.

Chitnis, M., Sorrell, S., Druckman, A., Firth, S.K. and Jackson, T. 2014. Who rebounds most? Estimating direct and indirect rebound effects for different UK socioeconomic groups. *Ecological Economics*, 106, 12–32.

Christoff, P. 1996. Ecological modernisation, ecological modernities. *Environmental Politics*, 5, 476–500.

Christoff, P. 2013. Transition state or state of emergency? Presented at the ECPR Workshop, Mainz.

Christoff, P. and Eckersley, R. 2011. Comparing state responses. In J.S. Dryzek, R.B. Norgaard and D. Schlosberg (eds), *The Oxford Handbook of Climate Change and Society*, pp. 431–448. Oxford University Press, Oxford.

Christoff, P. and Eckersley, R. 2013. *Globalization and the Environment.* Rowman & Littlefield, Plymouth, UK.

Chung, J.W. and Meltzer, D.O. 2009. Estimate of the carbon footprint of the US health care sector. *JAMA*, 302(18), 1967–1972.

Committee on Climate Change. 2010. *The Fourth Carbon Budget: Reducing Emissions through the 2020s.* Committee on Climate Change, London.

Committee on Climate Change. 2013. *Reducing the UK's Carbon Footprint.* Committee on Climate Change, London.

Committee on Climate Change. 2014. *Meeting Carbon Budgets: 2014 Progress Report to Parliament.* Committee on Climate Change, London.

Committee on Climate Change. 2015. *Reducing Emissions and Preparing for Climate Change – 2015 Progress Report to Parliament.* Committee on Climate Change, London.

Committee on Climate Change. 2016. *Meeting Carbon Budgets: 2016 Progress Report to Parliament.* Committee on Climate Change, London.

Concialdi, P. 2017. What does it mean to be rich? *Documents de travail de l'IRES*, 1.

Cook, S., Smith, K. and Utting, P. 2012. *Green Economy or Green Society? Contestation and Policies for a Fair Transition.* UNRISD Occasional Paper No. 10: Social Dimensions of Green Economy and Sustainable Development. United Nations Research Institute for Social Development, Geneva.

Coole, D. 2013. Too many bodies? The return and disavowal of the population question. *Environmental Politics*, 22, 195–215.

Coote, A. 2011. *The Great Transition: Social Justice and the Core Economy.* New Economics Foundation, London.

Coote, A. 2012a. *The Wisdom of Prevention: Long-Term Planning, Upstream Investment and Early Action to Prevent Harm.* New Economics Foundation, London.

Coote, A. 2012b. *Beyond Beveridge.* NEF Briefing. New Economics Foundation, London.

Coote, A. 2015. People, planet, power: Toward a new social settlement. *International Journal of Social Quality*, 5, 8–34.

Coote, A. and Franklin, J. 2009. *Green Well Fair: Three Economies for Social Justice.* New Economics Foundation, London.

Coote, A. and Franklin, J. (eds). 2013. *Time on Our Side: Why We All Need a Shorter Working Week.* New Economics Foundation, London.

Coote, A. and Harris, M. 2013. *The Prevention Papers: Upstream Investment and Early Action to Prevent Harm: Building Knowledge and Breaking Down Barriers.* New Economics Foundation, London.

Coote, A., Franklin, J. and Simms, A. 2010. *21 Hours: Why a Shorter Working Week Can Help Us All to Flourish in the 21st Century*. New Economics Foundation, London.

Corvalan, C., Hales, S. and McMichael, A. 2005. *Millennium Ecosystem Assessment: Ecosystems and Human Well-Being: Health Synthesis*. World Health Organization, Geneva.

Costanza, R. 1991. *Ecological Economics: The Science and Management of Sustainability*. Columbia University Press, New York.

Costanza, R., Daly, L., Fioramonti, L., Giovannini, E., Kubiszewski, I., Mortensen, L.F., Pickett, K.E., Ragnarsdottir, K.V., De Vogli, R. and Wilkinson, R. 2016. Modelling and measuring sustainable wellbeing in connection with the UN Sustainable Development Goals. *Ecological Economics*, 130, 350–355.

Crosland, A. 1956. *The Future of Socialism*. Jonathan Cape, London.

Crouch, C. 2011. *The Strange Non-Death of Neoliberalism*. Polity Press, Cambridge.

Crouch, C. 2015. *Governing Social Risks in Post-Crisis Europe*. Edward Elgar Publishing, Cheltenham, UK and Northampton, MA, USA.

Dahl, R.A. 1998. *On Democracy*. Yale University Press, New Haven, CT.

Daly, H.E. 1977. *Steady-State Economics: The Economics of Biophysical Equilibrium and Moral Growth*. W.H. Freeman, San Francisco, CA.

Daly, H.E. 1996. *Beyond Growth: The Economics of Sustainable Development*. Beacon Press, Boston, MA.

Daly, H.E. 2007. *Ecological Economics and Sustainable Development*. Edward Elgar Publishing, Cheltenham, UK and Northampton, MA, USA.

Davis, A., Hirsch, D. and Padley, M. 2014. *A Minimum Income Standard for the UK in 2014*. Loughborough University, Loughborough and Joseph Rowntree Foundation, York.

Davis, A., Hirsch, D., Padley, M. and Marshall, L. 2015. *How Much Is Enough? Reaching Social Consensus on Minimum Household Needs*. Centre for Research in Social Policy, Loughborough University, Loughborough.

Day, R., Walker, G. and Simcock, N. 2016. Conceptualising energy use and energy poverty using a capabilities framework. *Energy Policy*, 93, 255–264.

Deacon, B. 2007. *Global Social Policy and Governance*. Sage, London.

DECC. 2010. *Estimated Impacts of Energy and Climate Change Policies on Energy Prices and Bills*. Department of Energy and Climate Change, London.

DECC. 2013. *Guidance: Participating in the EU ETS*. Department of Energy and Climate Change, London.

Deeming, C. 2011. Determining minimum standards of living and household budgets: Methodological issues. *Journal of Sociology*, 47, 17–34.

del Rio, P. and Howlett, M. 2013. *Beyond the 'Tinbergen Rule' in Policy Design: Matching Tools and Goals in Policy Portfolios*. SSRN Scholarly Paper. SSRN, Rochester, NY.

Demailly, D., Chancel, L., Waisman, H. and Guivarch, C. 2013. *A Post-Growth Society for the 21st Century: Does Prosperity Have to Wait for the Return of Economic Growth?* IDDRI Study No. 08/13. Institut du développement durable et des relations internationales, Paris.

Demaria, F., Schneider, F., Sekulova, F. and Martinez-Alier, J. 2013. What is degrowth? From an activist slogan to a social movement. *Environmental Values*, 22, 191–215.

Department for Environment, Food and Rural Affairs. 2008. *A Framework for Pro-Environmental Behaviours*. Department for Environment, Food and Rural Affairs, London.

Dercon, S. 2012. *Is Green Growth Good for the Poor?* Policy Research Working Paper No. 6231. World Bank, Washington, DC.

De-Shalit, A. 2005. *Why Posterity Matters: Environmental Policies and Future Generations*. Routledge, Abingdon.

Dietz, S. 2011. From efficiency to justice: Utility as the informational basis of climate strategies. In J.S. Dryzek, R.B. Norgaard and D. Schlosberg (eds), *Oxford Handbook of Climate Change and Society*, pp. 295–308. Oxford University Press, Oxford.

Dietz, T., Rosa, E.A. and York, R. 2009. Environmentally efficient well-being: Rethinking sustainability as the relationship between human well-being and environmental impacts. *Human Ecology Review*, 16, 114–123.

Dobson, A. 1998. *Justice and the Environment*. Oxford University Press, Oxford.

Dobson, A. 2011. *Sustainability Citizenship*. Green House, Weymouth.

Dobson, A. and Bell, D. 2006. *Environmental Citizenship*. MIT Press, Cambridge, MA.

Dover, M.A. 2016. Human needs: Overview. In C. Franklin (ed.), *The Encyclopedia of Social Work*. Oxford University Press and National Association of Social Workers, New York.

Doyal, L. 1995. Needs, rights, and equity: More quality in healthcare rationing. *Quality in Health Care*, 4(4), 273–283.

Doyal, L. and Gough, I. 1991. *A Theory of Human Need*. Palgrave Macmillan, Basingstoke.

Druckman, Angela. 2016. Understanding households as drivers of carbon emissions. In Roland Clift and Angela Druckman (eds), *Taking Stock of Industrial Ecology*. Springer, Cham.

Druckman, A. and Jackson, T. 2008. Household energy consumption in the UK: A highly geographically and socio-economically disaggregated model. *Energy Policy*, 36, 3177–3192.

Druckman, A. and Jackson, T. 2009. The carbon footprint of UK households 1990–2004: A socio-economically disaggregated, quasi-multi-regional input–output model. *Ecological Economics*, 68, 2066–2077.

Druckman, A. and Jackson, T. 2010. The bare necessities: How much household carbon do we really need? *Ecological Economics*, 69, 1794–1804.

Druckman, Angela, Hartfree, Y., Hirsch, D. and Perren, K. 2011. *Sustainable Income Standards: Towards a Greener Minimum?* Joseph Rowntree Foundation, York.

Druckman, A., Buck, I., Hayward, B. and Jackson, T. 2012. Time, gender and carbon: A study of the carbon implications of British adults' use of time. *Ecological Economics*, 84, 153–163.

Dryzek, J.S. 2016. Institutions for the Anthropocene: Governance in a changing Earth system. *British Journal of Political Science*, 46(4), 937–956.

Dryzek, J.S., Downes, D., Hunold, C., Schlosberg, D. and Hernes, H.-K. 2003. *Green States and Social Movements: Environmentalism in the United States, United Kingdom, Germany, and Norway*. Oxford University Press, Oxford.

Duesenberry, J. 1960. Comment on 'An economic analysis of fertility'. In National Bureau of Economic Research, *Demographic and Economic Change in Developed Countries*, pp. 231–234. National Bureau of Economic Research, New York.

Duit, A. 2014. *State and Environment: The Comparative Study of Environmental Governance*. MIT Press, Cambridge, MA.

Duit, A., Feindt, P.H. and Meadowcroft, J. 2016. Greening Leviathan: The rise of the environmental state? *Environmental Politics*, 25, 1–23.

Duménil, G. and Lévy, D. 2011. *The Crisis of Neoliberalism*. Harvard University Press, Cambridge, MA.

Duro, J.A. and Teixidó-Figueras, J. 2013. Ecological footprint inequality across countries: The role of environment intensity, income and interaction effects. *Ecological Economics*, 93, 34–41.

Easterlin, R.A. 2001. Income and happiness: Towards a unified theory. *Economic Journal*, 111, 465–484.

Eckersley, R. 2004. *The Green State: Rethinking Democracy and Sovereignty*. MIT Press, Cambridge, MA.

Eckersley, R. 2015. The common but differentiated responsibilities of states to assist and receive 'climate refugees'. *European Journal of Political Theory*, 14, 481–500.

Eckersley, R. 2016. National identities, international roles, and the legitima-
    tion of climate leadership: Germany and Norway compared. *Environmental
    Politics*, 25, 180–201.
Ehrlich, P.R. and Holdren, J.P. 1971. Impact of population growth.
    Science, 171, 1212–1217.
Ekins, P. 2014. Strong sustainability and critical natural capital. In
    G. Atkinson, S. Dietz, E. Neumayer and M. Agarwala (eds), *Handbook of
    Sustainable Development*, 2nd edn, pp. 55–71. Edward Elgar Publishing,
    Cheltenham, UK and Northampton, MA, USA.
Ellis, G. and Barry, J. 2015. Low carbon transitions and post-fossil fuel
    energy transformations as political struggles: Analysing and overcoming
    carbon 'lock in'. *Academia.edu*. Available at: https://www.academia.edu/
    23407456/Low_Carbon_Transitions_and_Post-Fossil_Fuel_Energy_Tr
    ansformations_as_Political_Struggles_Analysing_and_Overcoming_Ca
    rbon_Lock_In.
Elson, D. 1988. Market socialism or socialization of the market? *New Left
    Review*, I, 3–44.
Environmental Audit Committee. 2008. *Personal Carbon Trading: Fifth
    Report of 2007–08*. House of Commons, London.
Environmental Tax Policy Institute and the Vermont Journal of
    Environmental Law. 2008. *The Reality of Carbon Taxes in the 21st
    Century*. Environmental Tax Policy Institute and the Vermont Journal
    of Environmental Law, South Royalton, VT.
Esping-Andersen, G. 1990. *The Three Worlds of Welfare Capitalism*.
    Princeton University Press, Princeton, NJ.
Esping-Andersen, G. 1999. *Social Foundations of Postindustrial Economies*.
    Oxford University Press, Oxford.
Eurostat. 2016. *Greenhouse Gas Emissions by Industries and Households*.
    Eurostat, Luxembourg.
Fankhauser, S. 2012. *A Practitioner's Guide to a Low-Carbon Economy:
    Lessons from the UK*. Policy Paper. Grantham Research Institute on
    Climate Change and the Environment, London.
Fankhauser, S. 2014. *Domestic Dynamics and International Influence:
    What Explains the Passage of Climate Change Legislation?* Grantham
    Research Institute on Climate Change and the Environment, London.
Farley, J., Schmitt Filho, A., Burke, M. and Farr, M. 2015. Extending
    market allocation to ecosystem services: Moral and practical impli-
    cations on a full and unequal planet. *Ecological Economics*, 117,
    244–252.
Farmer, J.D., Hepburn, C., Mealy, P. and Teytelboym, A. 2015. A third
    wave in the economics of climate change. *Environmental and Resource
    Economics*, 62(2), 329–357.

Fawcett, T. and Parag, Y. 2010. An introduction to personal carbon trading. *Climate Policy*, 10, 329–338.

Fell, D. 2016. *Bad Habits, Hard Choices: Using the Tax System to Make Us Healthier*. London Publishing Partnership, London.

Ferragina, E., Seeleib-Kaiser, M. and Spreckelsen, T. 2015. The four worlds of 'welfare reality' – social risks and outcomes in Europe. *Social Policy and Society*, 14, 287–307.

Fitzpatrick, T. 2011. Introduction. In T. Fitzpatrick (ed.), *Understanding the Environment and Social Policy*. Policy Press, Bristol, UK.

Fitzpatrick, T. 2014a. *Climate Change and Poverty: A New Agenda for Developed Nations*. Policy Press, Bristol, UK.

Fitzpatrick, T. 2014b. *International Handbook on Social Policy and the Environment*. Edward Elgar Publishing, Cheltenham, UK and Northampton, MA, USA.

Foresight. 2011. *Migration and Global Environmental Change: Future Challenges and Opportunities*. Government Office for Science, London.

Foster, J.B. 2009. *The Ecological Revolution*. Monthly Review Press, New York.

Frank, R.H. 2011. *The Darwin Economy: Liberty, Competition, and the Common Good*. Princeton University Press, Princeton, NJ.

Frankfurt, H. 1987. Equality as a moral ideal. *Ethics*, 98, 21–43.

Fraser, N. 2014. Behind Marx's hidden abode. *New Left Review*, 86, 55–72.

Freeman, R. 1992. The idea of prevention: A critical review. In S.J. Scott, G.H. Williams, S.D. Platt and H.A. Thomas (eds), *Private Risks and Public Dangers*. Avebury, Aldershot.

Freeman, R. 1999. Recursive politics: Prevention, modernity and social systems. *Children and Society*, 13, 232–241.

Frenken, K. 2017. Political economies and environmental futures for the sharing economy. *Philosophical Transactions of the Royal Society A*, 375, 20160367.

Frohlich, N. and Oppenheimer, J.A. 1992. *Choosing Justice: An Experimental Approach to Ethical Theory*. University of California Press, Berkeley.

Fuchs, D., Di Giulio, A., Glaab, K., Lorek, S., Maniates, M., Princen, T. and Røpke, I. 2016. Power: The missing element in sustainable consumption and absolute reductions research and action. *Journal of Cleaner Production*, 132, 298–307.

Fukuda-Parr, S. 2015. The 2030 Agenda and the SDGs – a course correction. *African Agenda*, 18, 5–6.

Funtowicz, S. and Ravetz, J. 1994. Uncertainty, complexity and post-normal science. *Environmental Toxicology and Chemistry*, 13, 1881–1885.

Gamble, A. 1995. The new political economy. *Political Studies*, 43, 516–530.

Gasper, D. 2007. Conceptualising human needs and wellbeing. In I. Gough and J.A. McGregor (eds), *Wellbeing in Developing Countries: From Theory to Research*. Cambridge University Press, Cambridge.

Gasper, D. 2012. Climate change: The need for a human rights agenda within a framework of shared human security. *Social Research*, 79, 983–1014.

Gassebner, M., Gaston, N. and Lamla, M.J. 2008. Relief for the environment? The importance of an increasingly unimportant industrial sector. *Economic Inquiry*, 46, 160–178.

Geels, F.W. 2014. Regime resistance against low-carbon transitions: Introducing politics and power into the multi-level perspective. *Theory, Culture and Society*, 31, 21–40.

Giddens, A. 1984. *The Constitution of Society*. Polity Press, Cambridge.

Giulio, A. and Fuchs, D. 2014. Sustainable consumption corridors: Concept, objections, and responses. *Gaia*, 23, 184–192.

Glyn, A. 2006. *Capitalism Unleashed: Finance, Globalization, and Welfare*. Oxford University Press, Oxford.

Goldemberg, J., Johansson, T.B., Reddy, A.K. and Williams, R.H. 1985. Basic needs and much more with one kilowatt per capita. *Ambio*, 14, 190–200.

Goodin, R.E. 1992. Towards a minimally presumptuous social welfare policy. In P. Van Parijs (ed.), *Arguing for Basic Income*, pp. 195–214. Verso, London.

Goodin, R.E., Headey, B., Muffels, R. and Dirven, H.J. 1999. *The Real Worlds of Welfare Capitalism*. Cambridge University Press, Cambridge.

Goodin, R.E., Rice, J.M., Parpo, A. and Eriksson, L. 2008. *Discretionary Time: A New Measure of Freedom*. Cambridge University Press, Cambridge.

Gordon, R.J. 2012. *Is U.S. Economic Growth Over? Faltering Innovation Confronts the Six Headwinds*. Working Paper No. 18315. National Bureau of Economic Research, Cambridge, MA.

Gore, T. 2015. *Extreme Carbon Inequality: Why the Paris Climate Deal Must Put the Poorest, Lowest Emitting and Most Vulnerable People First*. Oxfam International, Oxford.

Gorz, A. 1982. *Farewell to the Working Class: An Essay on Post-Industrial Socialism*. Pluto Press, London.

Gough, I. 1979. *The Political Economy of the Welfare State*. Critical Texts in Social Work and the Welfare State. Palgrave Macmillan, London.

Gough, I. 1994. Economic institutions and the satisfaction of human needs. *Journal of Economic Issues*, 28, 25–66.

Gough, I. 1996. Social welfare and competitiveness. *New Political Economy*, 1, 209–232.

Gough, I. 2000. *Global Capital, Human Needs and Social Policies: Selected Essays, 1994–99*. Palgrave Macmillan, Basingstoke.

Gough, I. 2008. European welfare states: Explanations and lessons for developing countries. In A.A. Dani and A. de Haan (eds), *Inclusive States: Social Policy and Structural Inequalities*, pp. 39–72. World Bank Publications, Washington, DC.

Gough, I. 2011a. *Climate Change, Double Injustice and Social Policy: A Case Study of the United Kingdom*. UNRISD Occasional Paper No. 1: Social Dimensions of Green Economy and Sustainable Development. United Nations Research Institute for Social Development, Geneva.

Gough, I. 2011b. From financial crisis to fiscal crisis. In K. Farnsworth and Z.M. Irving (eds), *Social Policy in Challenging Times: Economic Crisis and Welfare Systems*, pp. 49–64. Policy Press, Bristol, UK.

Gough, I. 2013. Carbon mitigation policies, distributional dilemmas and social policies. *Journal of Social Policy*, 42, 191–213.

Gough, I. 2014. Lists and thresholds: Comparing the Doyal–Gough theory of human need with Nussbaum's capabilities approach. In F. Comim and M.C. Nussbaum (eds), *Capabilities, Gender, Equality: Towards Fundamental Entitlements*, pp. 357–381. Cambridge University Press, Cambridge.

Gough, I. 2015a. Climate change and sustainable welfare: The centrality of human needs. *Cambridge Journal of Economics*, 39, 1191–1214.

Gough, I. 2015b. The political economy of prevention. *British Journal of Political Science*, 45(2), 307–327.

Gough, I. 2016. Welfare states and environmental states: A comparative analysis. *Environmental Politics*, 25, 24–47.

Gough, I. and Meadowcroft, J. 2011. Decarbonizing the welfare state. In J.S. Dryzek, R.B. Norgaard and D. Schlosberg (eds), *Oxford Handbook of Climate Change and Society*, pp. 490–503. Oxford University Press, Oxford.

Gough, I. and Therborn, G. 2010. The global future of welfare states. In F.G. Castles, S. Leibfried, J. Lewis and H. Obinger (eds), *The Oxford Handbook of the Welfare State*, pp. 703–876. Oxford University Press, New York.

Gough, I., McGregor, I.A. and Camfield, L. 2007. Theorizing wellbeing in international development. In I. Gough and J.A. McGregor (eds), *Wellbeing in Developing Countries: From Theory to Research*, pp. 3–44. Cambridge University Press, Cambridge.

Gough, I., Meadowcroft, J., Dryzek, J., Gerhards, J., Lengfeld, H., Markandya, A. and Ortiz, R. 2008. Climate change and social policy: A symposium. *Journal of European Social Policy*, 18, 325–344.

Gough, I., Abdallah, S., Johnson, V., Ryan-Collins, J. and Smith, C. 2011. *The Distribution of Total Embodied Greenhouse Gas Emissions by Households in the UK, and Some Implications for Social Policy.* CASE Papers No. CASE/152. Centre for Analysis of Social Exclusion, London School of Economics and Political Science, London.

Government Office for Science. 2011. *International Dimensions of Climate Change: Final Project Report.* Government Office for Science, London.

Grubb, M., Hourcade, J.C. and Neuhoff, K. 2014. *Planetary Economics.* Routledge, Abingdon.

Grunewald, N., Klasen, S., Martínez-Zarzoso, I. and Muris, C. 2012. *Income Inequality and Carbon Emissions.* Discussion Paper No. 92. Courant Research Centre, University of Göttingen, Göttingen.

Guillen-Royo, M. 2016. *Sustainability and Wellbeing: Human-Scale Development in Practice.* Routledge, Abingdon.

Habermas, J. 1987. *Knowledge and Human Interests.* Polity Press, Cambridge.

Habermas, J. 1989. *The Structural Transformation of the Public Sphere.* MIT Press, Cambridge, MA.

Hacker, J.S. and Pierson, P. 2002. Business power and social policy: Employers and the formation of the American welfare state. *Politics and Society*, 30(2), 277–325.

Haines, A., Alleyne, G., Kickbusch, I. and Dora, C. 2012. From the Earth Summit to Rio+20: Integration of health and sustainable development. *Lancet*, 379(9832), 2189–2197.

Hall, P.A. 1993. Policy paradigms, social learning, and the state: The case of economic policymaking in Britain. *Comparative Politics*, 25, 275–296.

Hall, P.A. and Soskice, D. 2001. *Varieties of Capitalism: The Institutional Foundations of Comparative Advantage.* Oxford University Press, Oxford.

Hamilton, Clive. 2010. *Requiem for a Species: Why We Resist the Truth about Climate Change.* Earthscan, London.

Hay, C. and Wincott, D. 2012. *The Political Economy of European Welfare Capitalism.* Palgrave Macmillan, Basingstoke.

Hayden, A. 2013. Patterns and purpose of work-time reduction: A cross-national comparison. In Anna Coote and Jane Franklin (eds), *Time on Our Side: Why We All Need a Shorter Working Week.* New Economics Foundation, London.

Hecht, S. 2002. Sacred groves and sacrifice zones: Ideologies of conservation and development. Presented at the Inaugural Symposium of the Rock Ethics Institute, Pennsylvania State University.

Held, D. and Hervey, A. 2011. Democracy, climate change and global governance: Democratic agency and the policy menu ahead. In D. Held, A. Hervey and M. Theros (eds), *The Governance of Climate Change: Science, Economics, Politics and Ethics*, pp. 89–110. Polity Press, Cambridge.

Helm, D. 2008. Climate-change policy: Why has so little been achieved? *Oxford Review of Economic Policy*, 24, 211–238.

Helm, D. 2011. The sustainable borders of the state. *Oxford Review of Economic Policy*, 27, 517–535.

Hemerijck, A. 2012. *Changing Welfare States*. Oxford University Press, Oxford.

Hepburn, C. 2009. Carbon taxes, emissions trading and hybrid schemes. In D. Helm and C. Hepburn (eds), *The Economics and Politics of Climate Change*, pp. 365–384. Oxford University Press, Oxford.

Hertwich, E.G. and Peters, G.P. 2009. Carbon footprint of nations: A global, trade-linked analysis. *Environmental Science and Technology*, 43, 6414–6420.

Hills, J. 2012. *Getting the Measure of Fuel Poverty: Final Report of the Fuel Poverty Review*. CASE Report No. 72. Centre for Analysis of Social Exclusion, London School of Economics and Political Science, London.

Hines, C. 2017. *Progressive Protectionism: Taking Back Control*. Park House Press, UK.

Hirvilammi, T., Laakso, S., Lettenmeier, M. and Lähteenoja, S. 2013. Studying well-being and its environmental impacts: A case study of minimum income receivers in Finland. *Journal of Human Development and Capabilities*, 14, 134–154.

HM Government. 2009. *The UK Low Carbon Transition Plan: National Strategy for Climate and Energy*. TSO, Norwich.

Hodgson, G.M. 2006. What are institutions? *Journal of Economic Issues*, XL, 1–25.

Hodgson, G.M. 2013. *From Pleasure Machines to Moral Communities: An Evolutionary Economics without Homo Economicus*. University of Chicago Press, Chicago, IL.

Hodgson, G.M. 2016. *Conceptualizing Capitalism: Institutions, Evolution, Future*. University of Chicago Press, Chicago, IL.

Höpner, M., Petring, A., Seikel, D. and Werner, B. 2014. *Liberalization Policy: An Empirical Analysis of Economic and Social Interventions in Western Democracies*. WSI Discussion Paper 192. Wirtschafts- und Sozialwissenschaftliches Institut, Düsseldorf.

House of Commons Energy and Climate Change Committee. 2012. *Consumption-Based Emissions Reporting*. House of Commons, London.

Howell, R.A. 2012. Living with a carbon allowance: The experiences of carbon rationing action groups and implications for policy. *Energy Policy*, 41, 250–258.

Hulme, D. and Scott, J. 2010. The political economy of the MDGs: Retrospect and prospect for the world's biggest promise. *New Political Economy*, 15, 293–306.

Hüther, M. 2013. Verteilungswirkungen des EEG. *Zeitschrift für Energiewirtschaft*, 37.

IEA. 2014. *World Energy Investment Outlook 2014: Special Report.* International Energy Agency, Paris.

IMF. 2014. *Redistribution, Inequality, and Growth.* International Monetary Fund, Washington, DC.

IPCC. 2007. *IPCC Fourth Assessment Report.* Intergovernmental Panel on Climate Change, Geneva.

IPCC. 2013. Summary for policymakers. In T.F. Stocker, D. Qin, G.K. Plattner, M. Tignor, S.K. Allen, J. Boschung, A. Nauels, Y. Xia, V. Bex and P.M. Midgley (eds), *Climate Change 2013: Physical Science Basis. Contribution of Working Group I to the Fifth Assessment Report of the Intergovernmental Panel on Climate Change*, pp. 1–30. Cambridge University Press, Cambridge.

IPCC. 2014a. *Climate Change Report 2014: Synthesis Report. Contribution of Working Groups I, II and III to the Fifth Assessment Report of the Intergovernmental Panel for Climate Change.* Intergovernmental Panel on Climate Change, Geneva.

IPCC. 2014b. Summary for policymakers. In C.B. Field, V.R. Barros, D.J. Dokken, K.J. Mach, M.D. Mastrandrea, T.E. Bilir, M. Chatterjee, K.L. Ebi, Y.O. Estrada, R.C. Genova, B. Girma, E.S. Kissel, A.N. Levy, S. MacCracken, P.R. Mastrandrea and L.L. White (eds), *Climate Change 2014: Impacts, Adaptation, and Vulnerability. Part A: Global and Sectoral Aspects. Contribution of Working Group II to the Fifth Assessment Report of the Intergovernmental Panel on Climate Change*, pp. 1–32. Cambridge University Press, Cambridge.

IPCC. 2014c. Summary for policymakers. In O. Edenhofer, R. Pichs-Madruga, Y. Sokona, E. Farahani, S. Kadner, K. Seyboth, A. Adler, I. Baum, S. Brunner, P. Eickemeier, B. Kriemann, J. Savolainen, S. Schlomer, C. von Stechow, T. Zwickel and J.C. Minx (eds), *Climate Change 2014: Mitigation of Climate Change. Contribution of Working Group III to the Fifth Assessment Report of the Intergovernmental Panel on Climate Change*, pp. 1–33. Cambridge University Press, Cambridge.

Jackson, T. 2009. *Prosperity without Growth: Economics for a Finite Planet.* Earthscan, London.

Jackson, T. 2017. *Prosperity without Growth: Foundations for the Economy of Tomorrow*, 2nd edn. Routledge, London.

Jackson, T. and Marks, N. 1999. Consumption, sustainable welfare and human needs – with reference to UK expenditure patterns between 1954 and 1994. *Ecological Economics*, 28, 421–441.

Jackson, T. and Papathanasopoulou, E. 2008. Luxury or 'lock-in'? An exploration of unsustainable consumption in the UK: 1968 to 2000. *Ecological Economics*, 68, 80–95.

Jackson, T. and Victor, P. 2013. *Green Economy at Community Scale*. Metcalf Foundation, Toronto.

Jackson, T. and Victor, P.A. 2015. Does credit create a 'growth imperative'? A quasi-stationary economy with interest-bearing debt. *Ecological Economics*, 120, 32–48.

Jacobs, M. 1991. *The Green Economy: Environment, Sustainable Development and the Politics of the Future*. Pluto Press, London.

Jacobs, M. 1995. Sustainable development, capital substitution and economic humility: A response to Beckerman. *Environmental Values*, 4, 57–68.

Jacobs, M. 2012. *Green Growth: Economic Theory and Political Discourse*. Working Paper No. 92. Grantham Research Institute on Climate Change and the Environment, London.

Jacobs, M. 2015. *The Paris Agreement Is Highly Ambitious and Very Clever*. Grantham Research Institute on Climate Change and the Environment, London.

Jacobs, M. and Mazzucato, M. 2016. *Rethinking Capitalism: Economics and Policy for Sustainable and Inclusive Growth*. John Wiley & Sons, Chichester.

Jessop, B. 2012. Economic and ecological crises: Green new deals and no-growth economies. *Development*, 55, 17–24.

Johansson, H., Khan, J. and Hildingsson, R. 2016. Climate change and the welfare state. In M. Koch and O. Mont (eds), *Sustainability and the Political Economy of Welfare*, pp. 94–108. Routledge, Abingdon.

Jordan, A., Benson, D., Wurzel, R. and Zito, A. 2011. Policy instruments in practice. In J.S. Dryzek, R.B. Norgaard and D. Schlosberg (eds), *The Oxford Handbook of Climate Change and Society*, pp. 536–549. Oxford Handbooks in Politics and International Relations. Oxford University Press, Oxford.

Jorgenson, A.K. 2014. Economic development and the carbon intensity of human well-being. *Nature Climate Change*, 4(3), 186–189.

Jorgenson, A.K., Schor, J.B., Knight, K.W. and Huang, X. 2016. Domestic inequality and carbon emissions in comparative perspective. *Sociological Forum*, 31(S1), September, 770–786.

Kallis, G. 2011. In defence of degrowth. *Ecological Economics*, 70, 873–880.

Kallis, G. 2015. *The Degrowth Alternative*. Great Transition Initiative. Available at: http://www.greattransition.org/publication/the-degrowth-alternative.

Kallis, G., Kalush, M., O'Flynn, H., Rossiter, J. and Ashford, N. 2013. 'Friday off': Reducing working hours in Europe. *Sustainability*, 5, 1545–1567.

Kasser, T. 2011. Capitalism and autonomy. In V.I. Chirkov, R.M. Ryan and K.M. Sheldon (eds), *Human Autonomy in Cross-Cultural Context*. Springer Netherlands, Dordrecht.

Kaul, I., Schelling, T., Solow, R.M., Stern, N., Sterner, T. and Weitzman, M. 2009. Round table discussion: Economics and climate change – where do we stand and where do we go from here? In J.-P. Touffut (ed.), *Changing Climate, Changing Economy*, pp. 135–164. Edward Elgar Publishing, Cheltenham, UK and Northampton, MA, USA.

Kemeny, J. 1981. *The Myth of Home-Ownership: Private versus Public Choices in Housing Tenure*. Routledge, Abingdon.

Kerkhof, A.C., Benders, R.M.J. and Moll, H.C. 2009. Determinants of variation in household CO_2 emissions between and within countries. *Energy Policy*, 37, 1509–1517.

Klein, N. 2015. *This Changes Everything: Capitalism vs. the Climate*, 1st edn. Simon & Schuster Paperbacks, New York.

Knight, K.W. and Rosa, E.A. 2011. The environmental efficiency of well-being: A cross-national analysis. *Social Science Research*, 40, 931–949.

Knight, K.W., Rosa, E.A. and Schor, J.B. 2013. Could working less reduce pressures on the environment? A cross-national panel analysis of OECD countries, 1970–2007. *Global Environmental Change*, 23, 691–700.

Koch, M. 2012. *Capitalism and Climate Change: Theoretical Discussion, Historical Development and Policy Responses*. Palgrave Macmillan, Basingstoke.

Koch, M. 2013. Welfare after growth: Theoretical discussion and policy implications. *International Journal of Social Quality*, 3, 4–20.

Koch, M. and Fritz, M. 2014. Building the eco-social state: Do welfare regimes matter? *Journal of Social Policy*, 43, 679–703.

Koch, M. and Mont, O. (eds). 2016. *Sustainability and the Political Economy of Welfare*. Routledge, Abingdon.

Kolstad, C., Urama, K., Broome, J., Bruvoll, A., Cariño Olvera, M., Fullerton, D., Gollier, C., Hanemann, W.M., Hassan, R., Jotzo, F., Khan, M.R., Meyer, L. and Mundaca, L. 2014. Social, economic and ethical concepts and methods. In O. Edenhofer, R. Pichs-Madruga, Y. Sokona, E. Farahani, S. Kadner, K. Seyboth, A. Adler, I. Baum, S. Brunner, P. Eickemeier, B. Kriemann, J. Savolainen, S. Schlomer, C. von Stechow, T. Zwickel and J.C. Minx (eds), *Climate Change 2014: Mitigation of Climate Change. Contribution of Working Group III to the Fifth Assessment Report of the Intergovernmental Panel on Climate Change*, pp. 207–282. Cambridge University Press, Cambridge.

Kopatz, M., Spitzer, M. and Christanell, A. 2010. *Energiearmut: Stand der Forschung, nationale Programme und regionale Modellprojekte in Deutschland, Österreich und Großbritannien.* Wuppertal Papers No. 184. Wuppertal Institut, Wuppertal.

Krugman, P. 2014. Could fighting global warming be cheap and free? *New York Times*, 18 September.

Kubiszewski, I., Costanza, R., Franco, C., Lawn, P., Talberth, J., Jackson, T. and Aylmer, C. 2013. Beyond GDP: Measuring and achieving global genuine progress. *Ecological Economics*, 93, 57–68.

Kühner, S. 2015. What if we waited a little longer? In Z. Irving, M. Fenger and J. Hudson (eds), *Social Policy Review 27: Analysis and Debate in Social Policy, 2015*, pp. 199–224. Policy Press, Bristol, UK.

Kuhnimhof, T., Armoogum, J., Buehler, R., Dargay, J., Denstadli, J.M. and Yamamoto, T. 2012. Men shape a downward trend in car use among young adults – evidence from six industrialized countries. *Transport Reviews*, 32, 761–779.

Kumhof, M., Rancière, R. and Winant, P. 2015. Inequality, leverage, and crises. *American Economic Review*, 105, 1217–1245.

Lafferty, W. 2012. Governance for sustainable development: The impasse of dysfunctional democracy. In J. Meadowcroft, O. Langhelle and A. Ruud (eds), *Governance, Democracy and Sustainable Development: Moving beyond the Impasse*, pp. 297–339. Edward Elgar Publishing, Cheltenham, UK and Northampton, MA, USA.

Lafferty, W. and Hovden, E. 2003. Environmental policy integration: Towards an analytical framework. *Environmental Politics*, 12, 1–22.

Lamb, W.F. and Rao, N.D. 2015. Human development in a climate-constrained world: What the past says about the future. *Global Environmental Change*, 33, 14–22.

Lamb, W.F., Steinberger, J.K., Bows-Larkin, A., Peters, G.P., Roberts, J.T. and Wood, F.R. 2014. Transitions in pathways of human development and carbon emissions. *Environmental Research Letters*, 9, 014011.

Langford, M. 2016. Lost in transformation? The politics of the Sustainable Development Goals. *Ethics and International Affairs*, 30, 167–176.

Lansley, S. 2016. *A Sharing Economy: How Social Wealth Funds Can Reduce Inequality and Help Balance the Books.* Policy Press, Bristol, UK.

Latouche, S. 2009. *Farewell to Growth.* Polity Press, Cambridge.

Laurent, E. 2015. *Inequality as Pollution, Pollution as Inequality.* Stanford Center on Poverty and Inequality, Stanford, CA.

Layard, R. 2005. *Happiness: Lessons from a New Science.* Allen Lane, London.

Layard, R. 2011. *Happiness: Lessons from a New Science*, 2nd edn. Penguin, London.

Liddle, B. 2015. What are the carbon emissions elasticities for income and population? Bridging STIRPAT and EKC via robust heterogeneous panel estimates. *Global Environmental Change*, 31, 62–73.

Lindblom, C.E. 1977. *Politics and Markets: The World's Political Economic Systems*. Basic Books, New York.

Lomborg, B. 2010. *Smart Solutions to Climate Change: Comparing Costs and Benefits*. Cambridge University Press, Cambridge.

Lundqvist, L.J. 2001. Implementation from above: The ecology of power in Sweden's environmental governance. *Governance*, 14(3), 319–337.

Mack, J. and Lansley, S. 1985. *Poor Britain*. George Allen & Unwin, London.

Marmot, M. 2005. Social determinants of health inequalities. *Lancet*, 365, 1099–1104.

Marmot, M. 2013. *Health Inequalities in the EU*. European Commission, Luxembourg.

Marshall, T.H. 1950. *Citizenship and Social Class*. Cambridge University Press, Cambridge.

Martin, S. 2010. Climate change, migration, and governance. *Global Governance: A Review of Multilateralism and International Organizations*, 16(3), 397–414.

Marx, K. 1926. *Capital*, Vol. 1. Charles Kerr & Co., London.

Mattioli, G. 2016. Transport needs in a climate-constrained world: A novel framework to reconcile social and environmental sustainability in transport. *Energy Research and Social Science*, 18, 118–128.

Max-Neef, M. 1989. Human scale development: An option for the future. *Development Dialogue* (Uppsala), 1, 5–80.

Mayne, R. 2017. How carbon reduction can also reduce poverty and inequality. Blog. Available at: https://oxfamblogs.org/fp2p/reframing-climate-ch ange-how-carbon-reduction-can-also-reduce-poverty-and-inequality/.

Maza, C. 2016. Britain's green subsidy cuts spook investors. *Global Warming Policy Forum*, 24 February. Available at: https://www.thegwpf.com/bri tains-green-subsidy-cuts-spook-investors/.

Mazzucato, M. 2011. The entrepreneurial state. *Soundings*, 49, 131–142.

McKibben, B. 2012. Global warming's terrifying new math. *Rolling Stone*, 19 July.

McKibben, B. 2015. Falling short on climate in Paris. *New York Times*, 13 December.

McPherson, K., Brown, M., Coote, A., Marsh, T. and Lobstein, T. 2009. Social class and overweight: Modelling the anticipated effects of social inequality on health service treatment costs until 2025. Unpublished paper commissioned by the UK Sustainable Development Commission and New Economics Foundation for the Marmot Review Task Group 5.

Meade, J.E. 1964. *Efficiency, Equality and the Ownership of Property*, 1st edn. Routledge, Abingdon.

Meadowcroft, J. 2005. From welfare state to ecostate. In J. Barry and R. Eckersley (eds), *The State and the Global Ecological Crisis*, pp. 3–23. MIT Press, Cambridge, MA.

Meadowcroft, J. 2012. Greening the state? In P.F. Steinberg and S.D. VanDeveer (eds), *Comparative Environmental Politics: Theory, Practice, and Prospects*, pp. 63–88. MIT Press, Cambridge, MA.

Meadows, D.H., Meadows, D., Randers, J. and Behrens, W., III. 1972. *Limits to Growth: A Report to the Club of Rome*. Universe Books, New York.

Medeiros, M. 2006. The rich and the poor: The construction of an affluence line from the poverty line. *Social Indicators Research*, 78, 1–18.

Mellor, M. 1997. *Feminism and Ecology*. Polity Press, Cambridge.

Mestrum, F. 2016. *The Social Commons*. Gerakbudaya, Petaling Jaya.

Michaelson, J., Abdallah, S., Steuer, N., Thompson, S., Marks, N., Aked, J., Cordon, C. and Potts, R. 2009. *National Accounts of Well-Being: Bringing Real Wealth onto the Balance Sheet*. New Economics Foundation, London.

Michaelson, J., Mahony, S. and Schifferes, J. 2012. *Measuring Well-Being: A Guide for Practitioners*. New Economics Foundation, London.

Milanovic, B. 2013. Global income inequality in numbers: In history and now. *Global Policy*, 4(2), 198–208.

Mol, A.P.J. 2016. The environmental nation state in decline. *Environmental Politics*, 25, 48–68.

Mol, A.P.J., Sonnenfeld, D.A. and Spaargaren, G. (eds). 2009. *The Ecological Modernisation Reader: Environmental Reform in Theory and Practice*. Routledge, London.

Monbiot, G. 2015. Grand promises of Paris climate deal undermined by squalid retrenchments. *Guardian*, 12 December.

Moore, J.W. 2015. *Capitalism in the Web of Life: Ecology and the Accumulation of Capital*. Verso, New York.

Morel, N. 2012. *Towards a Social Investment Welfare State? Ideas, Policies and Challenges*. Policy Press, Bristol, UK.

Murray, C. 2016. *In Our Hands: A Plan to Replace the Welfare State*. Rowman & Littlefield, Lanham, MD.

Nakano, S., Okamura, A., Sakurai, N., Suzuki, M., Tojo, Y. and Yamano, N. 2009. *The Measurement of CO_2 Embodiments in International Trade*. OECD Science, Technology and Industry Working Papers. Organisation for Economic Co-operation and Development, Paris.

Nässén, J. 2015. Konsumtionens övergripande utveckling. In J. Larsson (ed.), *Hållbara konsumtionsmönster – analyser av maten, flyget och den totala konsumtionens klimatpåverkan idag och 2050.* Report No. 6653. Naturvårdsverket, Stockholm.

Nässén, J. and Larsson, J. 2015. Would shorter working time reduce greenhouse gas emissions? An analysis of time use and consumption in Swedish households. *Environment and Planning C: Government and Policy,* 33(4), 726–745.

National Audit Office. 2009. *European Union Emissions Trading Scheme: A Review.* National Audit Office, London.

NEF. 2008. *A Green New Deal.* New Economics Foundation, London.

Neumayer, E. 2011. *Sustainability and Inequality in Human Development.* UNDP-HDRP Occasional Paper. United Nations Development Programme, New York.

Neumayer, E. 2013. *Weak versus Strong Sustainability: Exploring the Limits of Two Opposing Paradigms,* 4th edn. Edward Elgar Publishing, Cheltenham, UK and Northampton, MA, USA.

Newell, P. and Paterson, M. 2010. *Climate Capitalism: Global Warming and the Transformation of the Global Economy.* Cambridge University Press, Cambridge.

Newhouse, D.L., Suarez-Becerra, P. and Evans, M. 2016. *New Estimates of Extreme Poverty for Children.* Policy Research Working Paper No. 7845. World Bank Group, Washington, DC.

Niemietz, K. 2010. Measuring poverty: Context-specific but not relative. *Journal of Public Policy,* 30, 241–262.

Nilsson, M. and Eckerberg, K. (eds). 2007. *Environmental Policy Integration in Practice: Shaping Institutions for Learning.* Earthscan, London.

Nussbaum, M.C. 1993. Non-relative virtues: An Aristotelian approach. In M.C. Nussbaum and A. Sen (eds), *The Quality of Life,* pp. 242–269. Clarendon Press, Oxford.

Nussbaum, M.C. 2000. *Women and Human Development: The Capabilities Approach.* Cambridge University Press, Cambridge.

Nussbaum, M.C. 2006. *Frontiers of Justice: Disability, Nationality, Species Membership.* Harvard University Press, Cambridge, MA.

Nussbaum, M.C. 2011. *Creating Capabilities: The Human Development Approach.* Harvard University Press, Cambridge, MA.

OECD. 2013. Effective tax rate overview by country. In *Taxing Energy Use: A Graphical Analysis.* Organisation for Economic Co-operation and Development, Paris.

O'Connor, J. 1997. *Natural Causes: Essays in Ecological Marxism.* Guilford Press, London.

Offer, A. 2006. *The Challenge of Affluence: Self-Control and Well-Being in the United States and Britain since 1950.* Oxford University Press, Oxford.

Okun, A. 1975. *Equality and Efficiency: The Big Trade-Off.* Brookings Institution, Washington, DC.

O'Neill, J. 2011. The overshadowing of needs. In F. Rauschmayer, I. Omann and J. Frühmann (eds), *Sustainable Development*, pp. 25–43. Routledge, London.

O'Neill, J. 2015. Sustainability. In D. Moellendorf and H. Widdows (eds), *The Routledge Handbook of Global Ethics*, pp. 401–415. Routledge, London.

O'Neill, J. 2016. Happiness, austerity and inequality. In H. Rosa and C. Henning (eds), *Good Life beyond Growth: Critical Perspectives.* Routledge, Abingdon.

Ostrom, E. 1990. *Governing the Commons: The Evolution of Institutions for Collective Action.* Cambridge University Press, Cambridge.

Ostrom, E. 1996. Crossing the great divide: Coproduction, synergy, and development. *World Development*, 24, 1073–1087.

Ostrom, E. 2009. *A Polycentric Approach for Coping with Climate Change.* Policy Research Working Paper No. 5095. World Bank, Washington, DC.

Özkaynak, B., Adaman, F. and Devine, P. 2012. The identity of ecological economics: Retrospects and prospects. *Cambridge Journal of Economics*, 36(5), 1123–1142.

Piachaud, D. 2016. Citizen's income: Rights and wrongs. Available at: http://sticerd.lse.ac.uk/dps/case/cp/casepaper200.pdf.

Pickett, K., Wilkinson, R. and de Vogli, R. 2014. 'Equality, Sustainability and Wellbeing', The EU's Fifth Project: Transitional Governance in the Service of Sustainable Societies, Francqui International Conference, Brussels, Palais des Académies, 8–9 May.

Pierson, P. 2001. *The New Politics of the Welfare State.* Oxford University Press, Oxford.

Piketty, T. 2014. *Capital in the Twenty-First Century.* Harvard University Press, Cambridge, MA.

Pirgmaier, E. 2017. The neoclassical Trojan horse of steady-state economics. *Ecological Economics*, 133, 52–61.

Pogge, T. and Horton, K. 2008. *Global Ethics: Seminal Essays.* Paragon House Publishers, St Paul, MN.

Pogge, T. and Moellendorf, D. 2008. *Global Justice: Seminal Essays.* Paragon House Publishers, St Paul, MN.

Polanyi, K. 1944. *The Great Transformation: The Political and Economic Origins of Our Time.* Beacon Press, Boston, MA.

Porritt, J. 2006. *Capitalism as if the World Matters.* Routledge, London.

Power, A. and Zulauf, M. 2011. *Cutting Carbon Costs: Learning from Germany's Energy Saving Program.* London School of Economics and Political Science, London.

Pullinger, M. 2013. The 'green life course' approach to designing working time policy. In Anna Coote and Jane Franklin (eds), *Time on Our Side: Why We All Need a Shorter Working Week*. New Economics Foundation, London.

Putnam, H. 2002. *The Collapse of the Fact/Value Dichotomy and Other Essays*. Harvard University Press, Cambridge, MA.

Ranger, N. and Ward, B. 2013. *Unburnable Carbon 2013: Wasted Capital and Stranded Assets*. Grantham Research Institute on Climate Change and the Environment, London.

Rao, N.D. and Baer, P. 2012. 'Decent living' emissions: A conceptual framework. *Sustainability*, 4, 656–681.

Rao, N.D. and Min, J. 2017. Decent living standards: Material prerequisites for human wellbeing. *Social Indicators Research*. doi:10.1007/s11205-017-1650-0.

Rao, N.D., Riahi, K. and Grubler, A. 2014. Climate impacts of poverty eradication. *Nature Climate Change*, 4(9), 749–751.

Ravallion, M., Heil, M. and Jalan, J. 2000. Carbon emissions and income inequality. *Oxford Economic Papers*, 52(4), 651–669.

Rawls, J. 1971. *A Theory of Justice*. Clarendon Press, Oxford.

Raworth, K. 2012. *A Safe and Just Space for Humanity: Can We Live within the Doughnut?* Oxfam Discussion Paper. Oxfam, Oxford.

Raworth, K. 2017. *Doughnut Economics: Seven ways to think like a 21st-century economist*. RH Business Books, London.

Reed, H. and Lansley, S. 2016. *Universal Basic Income: An Idea Whose Time Has Come?* Compass, London.

Reusser, D., Lissner, T., Pradhan, P., Holsten, A., Rybski, D. and Kropp, J.P. 2013. Relating climate compatible development and human livelihood. *Energy Procedia*, 40, 192–201.

Ridley, M. 2010. *The Rational Optimist: How Prosperity Evolves*. Fourth Estate, London.

Roberts, S. 2008. Energy, equity and the future of the fuel poor. *Energy Policy*, 36, 4471–4474.

Rockström, J., Steffen, W., Noone, K., Persson, Å., Chapin, F.S., Lambin, E.F., Lenton, T.M., Scheffer, M., Folke, C., Schellnhuber, H.J. and Nykvist, B. 2009. A safe operating space for humanity. *Nature*, 461(7263), 472–475.

Room, G. 2016. *Agile Actors on Complex Terrains: Transformative Realism and Public Policy*. Routledge, Abingdon.

Rosa, E.A. and Dietz, T. 2012. Human drivers of national greenhouse-gas emissions. *Nature Climate Change*, 2(8), 581–586.

Royal Society. 2010. *Climate Change: A Summary of the Science*. Royal Society, London.

Ruger, J.P. 2009. *Health and Social Justice*. Oxford University Press, Oxford.

Ryan, R.M. and Deci, E.L. 2001. On happiness and human potentials: A review of research on hedonic and eudaimonic well-being. *Annual Review of Psychology*, 52, 141–166.

Ryan, R.M. and Sapp, A.R. 2007. Basic psychological needs: A self-determination theory perspective on the promotion of wellness across development and cultures. In I. Gough and J.A. McGregor (eds), *Wellbeing in Developing Countries: From Theory to Research*, pp. 71–92. Cambridge University Press, Cambridge.

Sabel, C.F. and Zeitlin, J. 2008. Learning from difference: The new architecture of experimentalist governance in the EU. *European Law Journal*, 14(3), 271–327.

Schaffrin, A. 2014. The new social risks and opportunities of climate change. In T. Fitzpatrick (ed.), *International Handbook on Social Policy and the Environment*. Edward Elgar Publishing, Cheltenham, UK and Northampton, MA, USA.

Schaffrin, A., Sewerin, S. and Seubert, S. 2015. Toward a comparative measure of climate policy output. *Policy Studies Journal*, 43(2), 257–282.

Schneider, F., Kallis, G. and Martinez-Alier, J. 2010. Crisis or opportunity? Economic degrowth for social equity and ecological sustainability. Introduction to this special issue. *Journal of Cleaner Production*, 18, 511–518.

Schor, J.B. 1991. Global equity and environmental crisis: An argument for reducing working hours in the North. *World Development*, 19, 73–84.

Schor, J.B. 2005. Sustainable consumption and worktime reduction. *Journal of Industrial Ecology*, 9, 37–50.

Schröder, M. 2009. Integrating welfare and production typologies: How refinements of the varieties of capitalism approach call for a combination of welfare typologies. *Journal of Social Policy*, 38, 19–43.

Schröder, M., Ekins, P., Power, A., Zulauf, M. and Lowe, R. 2011. *The KFW Experience in the Reduction of Energy Use in and CO_2 Emissions from Buildings: Operation, Impacts and Lessons for the UK*. Working Paper. UCL Energy Institute and LSE Housing and Communities, London.

Schwartz, H. and Seabrooke, L. 2008. Varieties of residential capitalism in the international political economy: Old welfare states and the new politics of housing. *Comparative European Politics*, 6, 237–261.

Scoones, I., Leach, M. and Newell, P. 2015. *The Politics of Green Transformations*. Routledge, London.

Scrieciu, S.Ş., Barker, T. and Ackerman, F. 2013. Pushing the boundaries of climate economics: Critical issues to consider in climate policy analysis. *Ecological Economics*, 85, 155–165.

SDC. 2008. *Carbon Emissions from Schools*. Sustainable Development Commission, London.

SDC. 2010. *Sustainable Development: The Key to Tackling Health Inequalities*. Sustainable Development Commission, London.

Seery, E. 2014. *Working for the Many: Public Services Fight Inequality*. Oxfam International, Oxford.

Sen, A. 1985. *Commodities and Capabilities*. Elsevier Science Publishers, Oxford.

Seyfang, G. and Paavola, J. 2008. Inequality and sustainable consumption: Bridging the gaps. *Local Environment*, 13, 669–684.

Shove, E., Watson, M. and Pantzar, M. 2012. *The Dynamics of Social Practice*. Sage, London.

Shue, H. 1993. Subsistence emissions and luxury emissions. *Law Policy*, 15, 39–60.

Shue, H. 2014. *Climate Justice: Vulnerability and Protection*. Oxford University Press, Oxford.

Skelton, A.C. and Allwood, J.M. 2017. Questioning demand: A study of regretted purchases in Great Britain. *Ecological Economics*, 131, 499–509.

Skidelsky, R. 2009. *Keynes: The Return of the Master*. Allen Lane, London.

Skidelsky, R. and Skidelsky, E. 2013. *How Much Is Enough? Money and the Good Life*. Penguin, London.

Smelser, N.J. and Swedberg, R. (eds). 2010. *The Handbook of Economic Sociology*, 2nd edn. Princeton University Press, Princeton, NJ.

Smith, K.R., Woodward, A., Campbell-Lendrum, D., Chadee, D., Honda, Y., Liu, Q., Olwoch, J., Revich, B. and Sauerborn, R. 2014. Human health: Impacts, adaptation, and co-benefits. In C.B. Field, V.R. Barros, D.J. Dokken, K.J. Mach, M.D. Mastrandrea, T.E. Bilir, M. Chatterjee, K.L. Ebi, Y.O. Estrada, R.C. Genova, B. Girma, E.S. Kissel, A.N. Levy, S. MacCracken, P.R. Mastrandrea and L.L. White (eds), *Climate Change 2014: Impacts, Adaptation, and Vulnerability. Part A: Global and Sectoral Aspects. Contribution of Working Group II to the Fifth Assessment Report of the Intergovernmental Panel on Climate Change*, pp. 709–754. Cambridge University Press, Cambridge.

Sommerer, T. and Lim, S. 2016. The environmental state as a model for the world? An analysis of policy repertoires in 37 countries. *Environmental Politics*, 25, 92–115.

Soper, K. and Emmelin, M. 2016. Reconceptualizing prosperity. In M. Koch and O. Mont (eds), *Sustainability and the Political Economy of Welfare*, pp. 44–58. Routledge, Abingdon.

Sovacool, B.K. and Linnér, B.-O. 2016. *The Political Economy of Climate Change Adaptation*. Palgrave Macmillan, London.

Spash, C.L. 2012. Editorial: Green economy, red herring. *Environmental Values*, 21, 95–99.

Spash, C.L. 2016. This changes nothing: The Paris agreement to ignore reality. *Globalizations*, 13, 928–933.

Steckel, J.C., Kalkuhl, M. and Marschinski, R. 2010. *Should Carbon-Exporting Countries Strive for Consumption-Based Accounting in a Global Cap-and-Trade Regime?* Potsdam Institute for Climate Impact Research, Potsdam.

Steckel, J.C., Jakob, M., Marschinski, R. and Luderer, G. 2011. From carbonization to decarbonization? Past trends and future scenarios for China's CO_2 emissions. *Energy Policy*, 39(6), 3443–3455.

Steckel, J.C., Brecha, R.J., Jakob, M., Strefler, J. and Luderer, G. 2013. Development without energy? Assessing future scenarios of energy consumption in developing countries. *Ecological Economics*, 90, 53–67.

Steed, S. and Kersley, H. 2009. *A Bit Rich*. New Economics Foundation, London.

Steffen, W. 2011. A truly complex and diabolical policy problem. In J.S. Dryzek, R.B. Norgaard and D. Schlosberg (eds), *Oxford Handbook of Climate Change and Society*, pp. 21–37. Oxford University Press, Oxford.

Steinberger, J.K. and Roberts, J.T. 2010. From constraint to sufficiency: The decoupling of energy and carbon from human needs, 1975–2005. *Ecological Economics*, 70, 425–433.

Steinberger, J.K., Roberts, J.T., Peters, G.P. and Baiocchi, G. 2012. Pathways of human development and carbon emissions embodied in trade. *Nature Climate Change*, 2(2), 81–85.

Stern, N. 2007. *The Economics of Climate Change: The Stern Review*. Cambridge University Press, Cambridge.

Stern, N. 2015. *Why Are We Waiting? The Logic, Urgency, and Promise of Tackling Climate Change*. MIT Press, Cambridge, MA.

Steward, F. 2012. Transformative innovation policy to meet the challenge of climate change: Sociotechnical networks aligned with consumption and end-use as new transition arenas for a low-carbon society or green economy. *Technology Analysis and Strategic Management*, 24(4), 331–343.

Stiglitz, J. 2013. *The Price of Inequality*. Penguin, London.

Stoker, G. 2009. *Nudge, Nudge, Think, Think: Experimenting with Ways to Change Civic Behaviour*. Bloomsbury Academic, London.

Storms, B., Goedemé, T., Van den Bosch, K. and Devuyst, K. 2013. *Towards a Common Framework for Developing Cross-Nationally Comparable Reference Budgets in Europe*. Methodological Paper No. 13/02. ImPRovE, Antwerp.

Strange, S. 1988. *States and Markets*. Continuum, London.

Streeck, W. 2010. *E Pluribus Unum? Varieties and Commonalities of Capitalism.* Max Planck Institute for the Study of Societies, Cologne.

Streeck, W. 2014. *Buying Time: The Delayed Crisis of Democratic Capitalism.* Verso, London.

Streeck, W. 2016. *How Will Capitalism End? Essays on a Failing System.* Verso, London.

Sunstein, C. and Thaler, R. 2006. Preferences, paternalism and liberty. In S. Olsaretti (ed.), *Preferences and Well-Being*, pp. 233–264. Cambridge University Press, Cambridge.

Swank, D. 2002. *Global Capital, Political Institutions, and Policy Change in Developed Welfare States.* Cambridge University Press, New York.

Taylor-Gooby, P. 2013. *The Double Crisis of the Welfare State and What We Can Do about It.* Palgrave Macmillan, Basingstoke.

Taylor-Gooby, P. and Stoker, G. 2011. The Coalition programme: A new vision for Britain or politics as usual? *Political Quarterly*, 82, 4–15.

Teixidó-Figueras, J., Steinberger, J.K., Krausmann, F., Haberl, H., Wiedmann, T., Peters, G.P., Duro, J.A. and Kastner, T. 2016. International inequality of environmental pressures: Decomposition and comparative analysis. *Ecological Indicators*, 62, 163–173.

Thaler, R.H. and Sunstein, C.R. 2008. *Nudge: Improving Decisions about Health, Wealth and Happiness.* Yale University Press, New Haven, CT.

Titmuss, R.M. 1970. *The Gift Relationship: From Human Blood to Social Policy.* George Allen & Unwin, London.

Torry, M. 2016. *The Feasibility of Citizen's Income.* Palgrave Macmillan, New York.

Townsend, P. 1979. *Poverty in the United Kingdom: A Survey of Household Resources and Standards of Living.* Penguin, Harmondsworth.

Trade Unions for Energy Democracy. 2012. *Resist, Reclaim, Restructure: Unions and the Struggle for Energy Democracy.* Rosa Luxemburg Stiftung, New York.

Turner, A. 2012. *Economics after the Crisis: Objectives and Means.* MIT Press, Cambridge, MA.

Turnheim, B., Berkhout, F., Geels, F., Hof, A., McMeekin, A., Nykvist, B. and van Vuuren, D. 2015. Evaluating sustainability transitions pathways: Bridging analytical approaches to address governance challenges. *Global Environmental Change*, 35, 239–253.

Ulvila, M. and Wilen, K. forthcoming. Engaging with the Plutocene: Moving towards degrowth and post-capitalist futures. In P. Heikkurinen (ed.), *Peaceful Coexistence for the Anthropocene*. Routledge, Abingdon.

UNDP. 2011. *Human Development Report 2011: Sustainability and Equity: A Better Future for All.* United Nations Development Programme, New York.

UNFCCC. 2015. *Report of the Conference of the Parties on Its Twenty-First Session, Held in Paris from 30 November to 13 December 2015*. United Nations Framework Convention on Climate Change, Paris.

UN General Assembly. 1948. Universal Declaration of Human Rights, 247 A (III).

Urry, J. 2011. *Climate Change and Society*. Polity Press, Cambridge.

van den Bergh, J.C. 2011. Environment versus growth – a criticism of 'degrowth' and a plea for 'a-growth'. *Ecological Economics*, 70, 881–890.

van den Bergh, J.C. and Verbruggen, H. 1999. Spatial sustainability, trade and indicators: An evaluation of the 'ecological footprint'. *Ecological Economics*, 29, 61–72.

Van Parijs, P. 1995. *Real Freedom for All. What (if Anything) Is Wrong with Capitalism?* Oxford University Press, Oxford.

Venugopal, R. 2015. Neoliberalism as concept. *Economy and Society*, 44(2), 165–187.

Verbist, G., Förster, M.F. and Vaalavuo, M. 2012. *The Impact of Publicly Provided Services on the Distribution of Resources: Review of New Results and Methods*. OECD Social, Employment and Migration Working Papers No. 130. OECD Publishing, Paris.

Victor, P.A. 2008. *Managing without Growth: Slower by Design, Not Disaster*. Edward Elgar Publishing, Cheltenham, UK and Northampton, MA, USA.

Victor, P.A. 2012. Growth, degrowth and climate change: A scenario analysis. *Ecological Economics*, 84, 206–212.

Vogler, J. 2007. The international politics of sustainable development. In G. Atkinson, S. Dietz and E. Neumayer (eds), *Handbook of Sustainable Development*, pp. 430–443. Edward Elgar Publishing, Cheltenham, UK and Northampton, MA, USA.

von Stechow, C., Minx, J.C., Riahi, K., Jewell, J., McCollum, D.L., Callaghan, M.W., Bertram, C., Luderer, G. and Baiocchi, G. 2016. 2 °C and SDGs: United they stand, divided they fall? *Environmental Research Letters*, 11(3), 034022.

Walker, G. and Burningham, K. 2011. Flood risk, vulnerability and environmental justice: Evidence and evaluation of inequality in a UK context. *Critical Social Policy*, 31, 216–240.

Warr, P. 1987. *Work, Unemployment and Mental Health*. Clarendon Press, Oxford.

Watts, N., Adger, W.N., Agnolucci, P., Blackstock, J., Byass, P., Cai, W., Chaytor, S., Colbourn, T., Collins, M., Cooper, A., Cox, P.M., Depledge, J., Drummond, P., Ekins, P., Galaz, V., Grace, D., Graham, H., Grubb, M., Haines, A., Hamilton, I., Hunter, A., Jiang, X., Li, M., Kelman, I., Liang, L., Lott, M., Lowe, R., Luo, Y., Mace, G., Maslin, M., Nilsson, M.,

Oreszczyn, T., Pye, S., Quinn, T., Svensdotter, M., Venevsky, S., Warner, K., Xu, B., Yang, J., Yin, Y., Yu, C., Zhang, Q., Gong, P., Montgomery, H. and Costello, A. 2015. Health and climate change: Policy responses to protect public health. *Lancet*, 386(10006), 1861–1914.

WBGU. 2009. *Solving the Climate Dilemma: The Budget Approach.* German Advisory Council on Global Change, Berlin.

WCED (World Commission on Environment and Development). 1987. *Our Common Future.* Brundtland Report. Oxford Paperbacks, Oxford University Press, Oxford.

Weber, C.L. and Matthews, H.S. 2008. Quantifying the global and distributional aspects of American household carbon footprint. *Ecological Economics*, 66, 379–391.

Weisbrod, B.A. and Hansen, W.L. 1968. An income–net worth approach to measuring economic welfare. *American Economic Review*, 58, 1315–1329.

Weitzman, M.L. 1974. *Is the Price System or Rationing More Effective in Getting a Commodity to Those Who Need It Most?* Working Paper No. 140. Massachusetts Institute of Technology (MIT), Department of Economics, Cambridge, MA.

Weitzman, M.L. 2009. On modeling and interpreting the economics of catastrophic climate change. *Review of Economics and Statistics*, 91(1), 1–19.

Whitmarsh, L. 2011. Social and psychological drivers of energy consumption behaviour and energy transitions. In S. Dietz, J. Michie and C. Oughton (eds), *The Political Economy of the Environment: An Interdisciplinary Approach*, pp. 213–228. Routledge, Abingdon.

Whitmee, S., Haines, A., Beyrer, C., Boltz, F., Capon, A.G., de Souza Dias, B.F., Ezeh, A., Frumkin, H., Gong, P., Head, P., Horton, R., Mace, G.M., Marten, R., Myers, S.S., Nishtar, S., Osofsky, S.A., Pattanayak, S.K., Pongsiri, M.J., Romanelli, C., Soucat, A., Vega, J. and Yach, D. 2015. Safeguarding human health in the Anthropocene epoch: Report of the Rockefeller Foundation–Lancet Commission on planetary health. *Lancet*, 386, 1973–2028.

WHO. 2015. *Health in 2015: From MDGs to SDGs.* World Health Organization, Geneva.

Wiggins, D. 1987. Essay 1: Claims of need. In *Needs, Values, Truth: Essays in the Philosophy of Value*, pp. 1–59. Oxford University Press, Oxford.

Wilkinson, R. and Pickett, K. 2009. *The Spirit Level: Why More Equal Societies Almost Always Do Better.* Allen Lane, London.

Williams, R. 1983. *Keywords: A Vocabulary of Culture and Society.* Fontana Press, London.

Wolf, C. 2009. Intergenerational justice, human needs, and climate policy. In A. Gosseries and L.H. Meyer (eds), *Intergenerational Justice*, pp. 347–376. Oxford University Press, Oxford.

Woll, C. 2014. *The Power of Inaction: Bank Bailouts in Comparison*. Cornell Studies in Political Economy. Cornell University Press, Ithaca, NY.

Woodcock, J., Edwards, P., Tonne, C., Armstrong, B.G., Ashiru, O., Banister, D., Beevers, S., Chalabi, Z., Chowdhury, Z., Cohen, A., Franco, O.H., Haines, A., Hickman, R., Lindsay, G., Mittal, I., Mohan, D., Tiwari, G., Woodward, A. and Roberts, I. 2009. Public health benefits of strategies to reduce greenhouse-gas emissions: Urban land transport. *Lancet*, 374, 1930–1943.

Woodward, D. 2015. Incrementum ad absurdum: Global growth, inequality and poverty eradication in a carbon-constrained world. *World Economic Review*, 4, 43–62.

Woodward, D. and Simms, A. 2006. *Growth Isn't Working*. New Economics Foundation, London.

World Bank. 2010. *World Development Report 2010: Development and Climate Change*. World Bank, Washington, DC.

World Bank. 2012. *Turn Down the Heat: Why a 4°C World Must Be Avoided*. World Bank, Washington, DC.

World Economic Forum. 2013. *The Green Investment Report*. World Economic Forum, Geneva.

York, R., Rosa, E.A. and Dietz, T. 2003. STIRPAT, IPAT and ImPACT: Analytic tools for unpacking the driving forces of environmental impacts. *Ecological Economics*, 46, 351–365.

Yu, E. and Liu, J. 2007. Environmental impacts of divorce. *Proceedings of the National Academy of Sciences*, 104, 20629–20634.

Zenghelis, Dimitri. 2012. *A Strategy for Restoring Confidence and Economic Growth through Green Investment and Innovation*. Centre for Climate Change Economics and Policy, Leeds and London.

Index